ROUTLEDGE LIBRARY EDITIONS: SCIENCE AND TECHNOLOGY IN THE NINETEENTH CENTURY

Volume 9

THE MERCURIAL CHEMIST

T0175114

THE MERCURIAL CHEMIST
A Life of Sir Humphry Davy

ANNE TRENEER

Routledge
Taylor & Francis Group

LONDON AND NEW YORK

First published in 1963 by Methuen & Co Ltd

This edition first published in 2019
by Routledge
2 Park Square, Milton Park, Abingdon, Oxon OX14 4RN

and by Routledge
52 Vanderbilt Avenue, New York, NY 10017

Routledge is an imprint of the Taylor & Francis Group, an informa business

British Library Cataloguing in Publication Data
A catalogue record for this book is available from the British Library

ISBN: 978-1-138-39006-5 (Set)
ISBN: 978-0-429-02175-6 (Set) (ebk)
ISBN: 978-0-367-02454-3 (Volume 9) (hbk)
ISBN: 978-0-367-02455-0 (Volume 9) (pbk)
ISBN: 978-0-429-39949-7 (Volume 9) (ebk)

Publisher's Note
The publisher has gone to great lengths to ensure the quality of this reprint but points out that some imperfections in the original copies may be apparent.

Disclaimer
The publisher has made every effort to trace copyright holders and would welcome correspondence from those they have been unable to trace.

The Mercurial Chemist

The Mercurial Chemist

A LIFE OF SIR HUMPHRY DAVY

by

ANNE TRENEER
M.A., B.Litt.

LONDON

METHUEN & CO LTD

36 ESSEX STREET WC2

First published 1963
© *1963 by Anne Treneer*
Printed in Great Britain
by Butler & Tanner Ltd
Frome & London
Cat No 2/2587/1

TO
MY FRIENDS
IN THE HUNDRED
OF PENWITH

Wisemen all wayes of knowledge past
To th' shepheards wonder come at last:

SIDNEY GODOLPHIN

Contents

Plates

Acknowledgements are due to the following
for permission to reproduce the illustrations:
Plate 1, the Royal Geological Society of
Cornwall; Plate 2, the City Art Gallery,
Bristol; Plates 3–6, the Royal Institution;
Plate 7, the Penzance Borough Museum;
Plate 8, Mrs M. E. Rolleston

Preface

This book had its origin in a paper, *Sir Humphry Davy and the Poets*, read before a small Cornish Society, and broadcast, in part, in the West Region. Davy's early life was as intensely and vividly local as his later life was cosmopolitan.

I came to have an abiding interest. From reading Davy as my taste directed in the successive volumes of his collected *Works*, which his brother edited, and which a friend gave me, I came to the early biographies, to later studies, and to pursuing Davy and Lady Davy through the memoirs and letters of their time. Davy was living in what he called the dawn of modern chemistry. His expositions are always lucid and often elegant. He sought unweariedly for unequivocal terms. He used no formulae. Towards the close of his life he sometimes contemplated writing the Memoirs of H. D., to which he would have given a title used by his favourite, Smollett, *The Adventures of an Atom*.

Many rhymes were made up about Davy in his lifetime. As late as 1905, E. C. Bentley wrote the well-remembered clerihew in *Biography for Beginners:*

> Sir Humphry Davy
> Abominated gravy.
> He lived in the odium
> Of having discovered sodium.

I have relied mainly on printed sources, as indicated in the Select Bibliography appended to this volume. In addition, I have been kindly permitted to examine manuscript material in the library of the Royal Institution in Albemarle Street. To the Managers acknowledgement is due for permission to print the draft of a poem by Davy on p. 4; drafts of letters on pp. 37 and 70; the letter from Davy to Coleridge on p. 60; Davy's parody of Wordsworth on p. 63, and other extracts from the notebooks as indicated in the text. The autograph letters from Coleridge to Davy, which I was allowed to read and copy from the originals in possession of the Royal Institution, are now included in the first four volumes of Professor Earl Leslie Grigg's great edition of Coleridge's Letters. From this edition, by courtesy of the Oxford University Press, I have quoted.

To Sir John Murray I am indebted for showing me letters in his possession, addressed to the John Murray of Davy's day, from Sir Humphry Davy, from Dr John Davy, and from Lady Davy. Acknowledgement is here made for permission to quote from Lady Davy's letter on p. 229.

Acknowledgements are due to Dr E. Weil and Messrs Taylor and Francis Ltd for permission to reproduce an extract from Dr Weil's paper published in *Annals of Science*, vol. 6, no. 3; to the Literary Trustees of Walter de la Mare, and to the Society of Authors as their representative, for permission to include the poem 'Napoleon' on p. 161; and to the Trustees of the Hardy Estate, and to Messrs Macmillan & Company Ltd, for permission to use as a tailpiece Thomas Hardy's poem, *The Youth Who Carried a Light*, taken from *The Collected Poems of Thomas Hardy*. The poem has no direct reference to Davy. It embodies what I should like to have been able to express.

I am very grateful to the many people who have answered my letters of inquiry; who have helped me with their special knowledge, or guided me to fresh material. I wish to thank the librarians of the County Library at Truro, and of the Penzance Library in the Morrab Garden at Penzance. To Mr K. D. Leach I am grateful for the information he was able to give me concerning Andrew Crosse, and for taking me to see the ruin of Fyne Court. I am obliged to the curator of the Royal Institution of Cornwall at Truro for showing me two interesting autograph letters from Davy to Davies Gilbert, other autograph letters of less interest, and Dr John Bingham Borlase's copy of the indenture which bound him 'to teach and instruct his apprentice [Humphry Davy] in all that appertained to the Art, Profession and Practice of a Surgeon and Apothecary, and not to use him in any kind of servile employment'. This legal and detailed covenant – Davy's copy is in the Penlee Museum at Penzance – is a reminder of the responsibilities undertaken by teacher, apprentice, and the guardian of the apprentice. Of Davy's mother it is written: 'the said Grace Davy doth hereby promise and agree to and with the said John Bingham Borlase that she, her Executors and Administrators, shall and will provide and find for the said Humphry Davy, during the term of five

years, sufficient Meat, Drink, washing and lodging and Apparrel of all
sorts fit and convenient for such an Apprentice'. Among other things,
Humphry bound himself to the injunction, 'that he shall not play at
any unlawful games whereby his said master may have any loss, nor
haunt Taverns and Playhouses, but in all things as a good and faithful
apprentice ought to do, shall and will demean and behave himself
towards his said Master, and all his during the said term'.

I should also like to express my warm thanks to Professor Herbert
Dingle for his friendly and helpful interest while this book was in
progress; and to Mr K. D. C. Vernon, librarian of the Royal Institution, for his help and advice. The parody of Wordsworth, of which he
sent me the transcript, was disembrangled, as Coleridge would have
said, from one of Davy's Clifton notebooks of 1800.

Too late to make use of the wider range of material indicated, I
read two essays. The first, which concerns John Davy, is entitled
'John Davy, Physician, Scientist, Author, Brother of Sir Humphry',
by Richard S. Ross (*Bulletin of the History of Medicine*, vol. XXVII,
no. 2, March–April, 1953). For lending me his reprint of this essay
I am indebted to Mr J. L. Rolleston, John Davy's great grandson, to
whom I am also grateful for information about his great grandfather's
portrait reproduced in the present volume.

Through Mr Rolleston I heard of manuscript and printed matter
relating chiefly to John Davy, but of interest also to students of his
brother, now in the Library of the University of Keele. It has not yet
been fully examined. I wish to thank the librarian for sending me an
outline of the collection, and for a photocopy of a copy of a letter by
John Davy to Eliza Fletcher describing his experience in Brussels as
a newly qualified doctor before and immediately after Waterloo. This
letter, with its emphasis on the agonizing scene of the arrival of the
wounded and the horrors in the hospital as the numbers continued to
increase, helps to explain why Mrs Davy's second son did not accept
his brother's invitation to join him in his later researches, but continued to practise as an army doctor while making his own contribution
to knowledge of the behaviour of human and animal bodies.

The second essay, 'Lady Davy in her Letters', by W. M. Parker,
M.B.E. (*The Quarterly Review*, January 1962), gives, through
copious quotation from her letters, a fairer impression of Lady Davy
than I have been able to set down. Among other good things is her
account, in a letter to Scott dated March 31st, 1825, of three days

spent with Goethe at Weimar. Her tour of Germany was made while Davy was in Scandinavia and North Germany. He must have been enjoying the 'social gaiety' of Gauss at about the time Lady Davy was talking with Goethe who, at 76, was 'full of vigour, of conversation, and courtesy of manner'. It seems a pity that Davy, who met so many poets, did not meet the Northern Apollo, as Lady Davy called him; for Goethe was not only aware and argumentative in the sciences but imaginatively quickened. Goethe's bearing towards Lady Davy confirms the impression one has from general reading, that Sir Humphry in his lifetime, and particularly on the Continent, was considered almost equal in renown with Scott and Byron, singled out by Goethe as his two most favourite authors.

ANNE TRENEER

1 · *Childhood and Schooldays*

'It is surely a pure delight . . .' Delight was one of Humphry Davy's favourite words, including for him elation of spirit, and a kind of hilarity in the use of his swift mind. His discoveries, lucid expositions, and discourses have their place in the history of science; his passion was to know. But we have sight of him, too, as he was in the circumstances of his time – a man of mercurial temperament and enthusiastic energy of character, whose flashes of insight, and zest for field sports, made him a friend of anglers and poets. Coleridge read *Christabel* aloud to him, and Wordsworth a part of *The Recluse* as he first planned it. When Dorothy Wordsworth wrote to Lady Beaumont that there was too much company at Dove Cottage for William to begin in earnest on his important Task, she added that it must not be supposed but that they had great joy at the sight of some who came. Such a man as Mr Davy, who had just paid them a brief visit, was, she said, a treasure anywhere. That was in 1805.

By 1826 poet and scientist, or chemical philosopher as Davy expressed it, had diverged, and there were fewer points of contact between them than might have been expected. Wordsworth had followed that way of insight into truth of which Coleridge said the previous condition was that a man should dare commune with his very and permanent self. Davy had been intent on the unfathomable working of things; on the method by which man might pass from received knowledge to tested knowledge previously hidden and, by ingeniously applying it, ease, as he hoped, labour and pain. Because wonder essentially underlay the two modes, Coleridge's imagination comprehended both. There was a time when he thought Wordsworth and Davy the two great men he knew, though he admitted he was given to romanticizing men's characters. For Davy's other favourite word was glory, glory in its liturgical use, Heaven and earth are full of Thy glory; but also meaning desire of fame and joy in success.

Davy was roughly eight years younger than Wordsworth and six years younger than Coleridge. He was born in Penzance on December 17th, 1778, and died at Geneva on the 29th May, 1829. These conditions of time and place made a difference to his view of the world, and in particular to his feeling about and judgment of the French Revolution; he missed the lyrical hope in it. Wordsworth first went to France in July 1790; Davy in October 1813. In his youth Wordsworth's theme had been not only what passed in himself but how, by virtue of poetry, to relate his love of nature to the love of man, and to the vision by which he had been enraptured in France before 'the sorrowful reverse for all mankind'. With Davy that dream was replaced by hope in what we have come to call the scientific and technological revolutions which he heralded. His life was not simple. Perhaps what he hoped contradicted what he was. He sought power through curiosity of wit, that sharpness of our understanding with which, mediaeval mystics warned, is mingled always some manner of fantasy; but he came, through his mother, from one who knew more of the power of love, not as a contemplative, but through the exertion called out by the hopes, wants, and wishes of a rising family and needy neighbours. He lived during the first onslaught of industrial ugliness; but he was nurtured on the beauty of the world as it was when Wordsworth was making it felt in poetry and Constable was painting it. Like them, he was brought up in a remote place and knew much solitude when he was young. There was a difference. I cannot find that there was either a miner or a deep-sea fisherman in Davy's ancestry; but he knew by closest proximity, in the intimate way of a curious child, these intense and dangerous toils and saw, almost before he could walk, how an engine could help.

Davy was a Cornishman. It would be easy to emphasize his 'Celtic' affinities. He was born and bred in west Cornwall, not leaving it until he was nearly twenty; Dr Beddoes, with whom he was associated at Clifton, was of Welsh extraction; he was a friend of the Anglo-Irish Edgeworths; he married a Scot who was the widow of a Welsh squire. By the time he was growing up Dr William Borlase had completed his learned work on Cornish Antiquities. Dolly Pentreath, one of several somewhat younger speakers of the ancient Cornish language, had died only the year before Humphry was born. His poems show that his imagination was kindled by Cornish stories. But to say that he thought of himself as Celtic, or differing radically from his fellow-Englishmen, would be to give a false impression. The whole bent of his mind was towards an international exchange of ideas, and away from all forms of sectarianism.

With this proviso we may notice how Cornish Davy was, both in his original gifts and his weaknesses, these inherent qualities being only the stronger in him because not explored by his own consciousness. His was an active nature. Once, when asked if he would not like to go on a whaling expedition, he said only if he could wield the harpoon himself, else it would be a spectacle, not an enterprise.

In his ancestry there is no name so resoundingly Cornish as Sophia Trevanion of St Michael Carhayes, Byron's grandmother, though his paternal grandmother lived for a time in Tregony, a parish adjoining Carhayes in south Cornwall. She lived in a haunted house about which she used to tell her small grandson. She was very composed with ghosts, thinking it more extraordinary that human beings should not see ghosts than that they should. Originally she was of the wild, religious, mining parish of St Just in Penwith, where the sound of the sea is everywhere heard, where in clear weather the air is so insubstantial that one feels as light as a mote, while on grey, misty, strange days the past crowds the present, so full the region seems of all that has passed away.

How the first Davys came to west Cornwall is not certain; we cannot claim them as intelligent mushrooms sprung out of the Cornish soil. But by the time Humphry was born they had, by continuous occupation of upwards of two hundred years, and by marriage with wives from the parishes in the neighbourhood of Mounts Bay, become settled inhabitants of the Hundred of Penwith. The name, in the varied spelling of a spell-by-ear age, appears in most of the early parish registers of West Penwith: in Madron and Paul; in St Hilary with Marazion; in Ludgvan, Gulval, Zennor, and St Just.

Humphry Davy's branch of the Davy family was of Ludgvan Parish. The name appears in the earliest parish registers. Tablets in the church commemorate Davy wives in the seventeenth and eighteenth centuries, both pious women, and one of 'sweet conversation', discreet, charitable, and loved by the poor who wept for her when she died. The men tended to be yeoman farmers of steady intelligence with substantial stones at their heads when they came to die. Seven of these may be seen in a row – they were placed against the church wall after an alteration to Ludgvan church and churchyard. Some stones carry initials only, in deep-cut lasting letters. The names of

Davy's grandparents are carved in full, Edmund Davy and Grace Davy.

That Davy looked at these tablets and stones, and wondered about his ancestors, is clear from a notebook in which is the draft of a poem on which he worked subsequently. The finish of the later poem removes us from the young Davy, groping towards a thought which never left him. The early poem is almost unpunctuated:

> My eye is wet with tears
> For I see the white stones
> > That are covered with names
> > The stones of my forefathers graves
>
> ———
>
> No grass grows upon them
> For deep in the earth
> > In darkness and silence the organs of life
> > To their primitive atoms return
>
> ———
>
> Through ages the air
> Has been moist with their blood
> Through ages the seeds of
> > the thistle has fed
> > On what was once motion and form
>
> ———
>
> The white land that floats
> Through the heavens
> Is pregnant with
> > that which was life
> And the moonbeams
> > that whiten it came
> From the breath and the spirit of man.
>
> ———

[4]

Thoughts roll not beneath the dust
No feeling is in the cold grave.
Neither thought nor feeling can die
They have leaped to other worlds
They are far above the skies

———

They kindle in the stars
They dance in the light of suns
Or they live in the comet's white haze

———

These poor remains of frame
Were the source of the organs of flesh
That feel the control of my will
That are active and mighty in me.

They gave to my body form
Is nought in your dying limbs
That gave to my spirit life
The blood that rolled through their veins
Was the germ of my bodily power.

Their spirit gave me no germ
of kindling energy . . .

The poem breaks off. In ordered print it loses something of the
power it has on the written page to make us share Davy's feeling.

Humphry's father, Robert Davy, the eldest son of Edmund and
Grace, was brought up in the house of an uncle, another Robert Davy,
who had considerable property in Ludgvan, and was the friend of
Dr William Borlase, Rector of Ludgvan, antiquary, historian, and a
friend of Pope's. Just as one can feel the kind of man Borlase was in
every sentence he wrote, and become attached to his druidical theories
against all later evidence, so one can almost feel what kind of a man his
good friend and neighbour Humphry's great-uncle, Robert Davy, must
have been. It is clear that the Davys were not unlettered or boorish
people, but neither is it on record that any one of them achieved

intellectual eminence of any kind. There is a hint that, like their descendant Humphry, they enjoyed good living. Writing to his wife after the onslaught of his own illness Humphry Davy said, 'My grandfather, and   of my great uncles died of Apoplexy'. In the space which I have left John Davy inserted the figures *six* or *seven*; but it looks more like *one* in the original letter preserved among other material in the Royal Institution in Albemarle Street. Davy adds, in the letter, that there was a case of the same kind in his mother's family. For eight or nine to have died of apoplexy would seem excessive even in that apoplectic age.

Although Humphry's father lived in his uncle's house, and was expected to be his heir, he was also a craftsman. He is best remembered as a woodcarver and gilder; but as his own father was a builder, it seems likely that he had also some skill in constructive carpentry, or at least in the furniture-making of the time which fitted the character of the woodwork in houses. He went to London to perfect his wood-carving. In the possession of the Victoria and Albert Museum is a chimney-piece front attributed to Robert Davy, with a carving of the fable of the fox and the stork in the centre compartment and scroll work at the sides. When Robert Davy returned to Penzance he carved chimney-pieces with skill and pleasure until carved chimney-pieces went out of fashion as moulding came in. An example of his art may be seen in Ludgvan Rectory. It is a chimney-piece elegant in design, with a spirited carving of the Ludgvan tin-stamp – two griffins with medallions. At Nancealverne, the home of the late Judge Scobel Armstrong, hangs a picture said to have been framed by Robert Davy. It is an oval of twined leaves, not florid, but free and bold – a *scherzo*. Although Robert Davy turned later to farming, Humphry was known locally as little carver Davy's boy because of his father's original occupation and small stature. He was shorter than either of his sons and they were not tall.

On his mother's side Davy came from the interesting family of Millett of St Just. His two grandmothers, both named Adams, though of different families, came from the same neighbourhood. His great-grandmother was a Usticke of Botallack, a family of ancient standing in Cornwall with a funeral hatchment in Madron Church. Both Davy's mother and his paternal grandmother must have been remarkable women, opposites, but combining in their two persons the qualities which have distinguished many Cornishwomen who have brought up able sons. Grace Millett, Humphry's mother, was a woman of fine fibre, possessing the virtues of deep piety, steadfastness, and courage; Grace Adams, his paternal grandmother, was of a more exalted temperament, imaginative, imbued with a feeling for the place in which she lived, treasuring old stories, practising old traditions, believing in ghosts and marvels, while acting always with great common-sense. It was she who told Humphry stories. From her he had, in part, his sense of the beauty of the world and the feeling, which never left him, of neighbouring the invisible. From his father, I think, he had his sanguine temperament, his rapidity and dexterity in manipulation, his love of sport, and his recklessness. The sense of decorum, respectability, and politeness imposed on his dancing faculty of mind came perhaps from the family of Tonkin. John Tonkin, a famous Penzance worthy, had a special place in the history of Humphry Davy. Although he was not a relation he brought up Humphry's mother, and in part Humphry himself.

As a young man John Tonkin, surgeon and apothecary, son of Uriah Tonkin and brother of the Reverend William Tonkin, lodged with the Milletts, Humphry Davy's maternal grandparents, who at that time kept a mercer's shop in Penzance. On the third of June 1757, Humphry Millett died; on the ninth his wife died of the same fever leaving three little girls. John Tonkin continued to live in the house. He invited the children's cousin, Peggy Adams, to come and look after them, and he acted as guardian to the three girls until they grew up and married. Jane married Henry Sampson, a watchmaker of Penzance; Elizabeth married her cousin, Leonard Millett, a gentleman of Marazion; Grace, in 1776, married Robert Davy.

Two years later, at five o'clock in the morning on the 17th December, their first child was born. The house was No. 4 The Terrace, in what is now called Market Jew Street. Then, the terrace was wider, and sloped gradually to a narrow cart-track – the coach road ended at Marazion. There were trees, including a handsome oriental plane. In later life Humphry imagined that his sense of the intimate association of life, light, and motion might have been due to very early recollection of wind playing in these trees which have long since gone. The house itself has been pulled down. The site, occupied by Messrs Oliver's shoe-shop, and marked by a plaque, is not far from where Davy's statue stands, and is on the right as one goes up Market Jew towards Land's End. The baby was christened in old St Mary's, then a chapel-of-ease to Madron Parish Church, the mother-church of Penzance. The entry in Madron Parish Register is: 'Humphry son of Robert Davy – baptized in Penzance Chapel 22 Jan. 1779.' He was given the Millett Christian name of Humphry. In the course of time four other children were born, Katherine (Kitty), Grace, Elizabeth (Betsy), and John, John being twelve years younger than Humphry.

In his youth Davy had a natural gladness. He was brought up amidst relatives of varied ages; loved, fostered, cheered, and clapped. When John Davy came to write his brother's biography it was still remembered in the family how forward Humphry had been, how he had walked off when he was only nine months old, was full of talk when he was two, could recite little prayers and stories before he could read, and managed to draw in great letters the names of the characters in *Aesop's Fables* and *The Pilgrim's Progress* – his favourite book when he was young, and one which he knew almost by heart – before he had learned to write. He made up verses at five, and recited them at the Christmas gambols. When he had learned to read, it was remarked that he could read nearly as fast as the pages would turn, an accomplishment common in quick children if they are not taught to read by methods too elaborate. Grown people who have never had the faculty, or who have lost it, always wonder at this natural speed of childhood. Davy retained it as a valuable asset all his life. He was beloved by aunts. In fact, like most other first babies in Cornwall which takes special notice of its spirited children, Humphry was the centre of affection and returned it. The desire to please was to remain with him, making him likable and vulnerable all his life. Much of his education was through talk. Fishermen were everywhere, and fishermen, who work strenuously at odd times and seasons, often seem to have more leisure than other folk. One has only to go to Mousehole now to see men pacing 'The Cliff' as though pacing a deck, and talking all day. If there are three together one will walk backwards facing his interlocutors, so as not to miss a word and be sure of driving his own word home.

Humphry talked, listened, looked and imitated the grown-ups who were engaged on all sorts of processes in or near Market Jew Street. At the foot of it towards Chyandour were tan-pits. He could watch people curing hides, making boots, and selling them on the open 'standings' by the Market House. He could see his uncle Sampson making clocks. He could follow the mules laden with copper coming from St Just and going down Chapel Street towards the Quay.

Childhood and Schooldays

The Terrace was raised above the level of steep Market Jew Street, down which the water ran in rainy weather. Humphry, imitating his father and his uncle Leonard, great anglers both, would stand and make casts with a piece of string and a bent pin. And then there were the quays at the Mount, at Penzance, at Newlyn and Mousehole. Today one may see a father or a grandfather on any of these quays, hand in hand with a child not much more than a baby, pausing by each interesting sight as long as the child chooses, and answering his questions, not in a bored or educational way. In shared companionable communion, talking they go. After the usual play with turnip-lanterns and fireworks Humphry came to love, above all things, angling. From fishing in the gutter of Market Jew he came to fishing from the rocks, and angling, after Izaak Walton's fashion, in the streams.

He was to care almost equally for sport with a gun; this taste, too, planted in him as a boy, grew as he grew. He lived where both sports were part of life. The Borough of Penzance, between Marazion and its elder neighbour, Mousehole, was compact and rural, although it was a coinage town, one of the towns of the Duchy to which tin was brought after its reduction, to be stamped before being exported. A case showing specimens of 'hot-marks' or smelter's trade marks from the local smelting-houses may be seen in the museum at Penlee House, Penzance. One stamp is of the Agnus Dei, a device which carries the mind back, like the inn sign, 'The Lamb and Flag', to the far-off days when the Knights Hospitallers had an establishment at Landithy, in Madron parish. Humphry, and later his brother John, would see, in well-arranged piles, the massive glittering blocks of stamped tin which encumbered the Market Place.

Davy's babyhood was in Penzance; but, while he was still very young, the family moved from the house in Market Jew to a copyhold property of seventy-nine acres in Ludgvan Parish, about two and a half miles from Penzance. Robert Davy had not become his uncle's heir, though John Davy says that a will unsigned was found in his favour. He had, however, this smaller place and there, at Varfell, he built a house with a glorious view of St Michael's Mount across the Marazion marshes.

It would be impossible to over-estimate the influence of his early environment on Davy. The Marazion marshes even now, in spite of the railway, have an extraordinary wild beauty. It is felt especially in winter, towards a still January sunset, when the pale gold rushes are reflected in the water, and the timeless notes of the wild fowl which haunt the pools are heard, together with the sound of their wings. Because of lucky folds in the earth it still seems from Varfell as though only flat fields, and the marsh, and a strange dark wood separate it from the sea and the Mount. The Mount was to Davy a visible symbol of romance; it is from nowhere more royally viewed than from the heights above Ludgvan. On a winter morning when the sun is low you may see it in a wash of silver light. Davy cared for it most by moonlight – it was under the full moon that he was most strongly to feel visitations, feelings of kinship with nature which he tried to express with all the resources of his youthful art. He saw into common things with the penetration and lucidity of childhood, and with a picturing faculty which he never lost. A generation before, William Oliver of Trevarnoe, the celebrated Bath physician, the friend of Pope and of William Borlase, had written of his own boyhood in Ludgvan parish: ' 'Tis not only with our own species that we contract the most lasting friendships in the beginning of life; I remember the name and character of every dog I used to miss school to hunt with; I could go to every little thicket which was most likely to afford game; I love the memory of a tall sycamore out of which I used to cut whistles. I have the situation of the hazel from which I obtained the best cob nuts full in my eye; and I remember with gratitude a rare (apple) tree, which afforded the first *regale* of the summer, and the Borlase's Pippin which, like its namesake, was a high entertainment in a winter's evening, in a warm room, with a good fire.' Humphry planted an apple-tree in his own garden at Varfell, and was to await impatiently the arrival of apples from it when at one time he lay ill in London. He had a pony, Derby, and a water spaniel, Chloe, one of a litter he had begged when it had been doomed to drowning in puppyhood – but this was later. On his pony he rode to and fro between Varfell and Penzance by a track which allows the Mount to be constantly in view.

Schooling did not interfere too much with his natural tendencies. He went first to the Writing School at Penzance, and then to the Latin School kept by the Rev. J. C. Coryton, who seems to have used a mixture of ferocity and playfulness. Davies Gilbert had been a pupil at the school under a former master. No strict records were kept of Master Davy's progress and misdemeanours. He seems to have lived as careless as a trout, and if Mr Coryton pulled his ears, he retaliated by appearing with his ears in plaster and replying, when asked the reason, that it was to prevent a mortification. The part of Harlequin which he took in a pantomime with the other boys is in character for him. No doubt Mr Coryton often had the best of the game, for the first rhyme on Davy's name was made by him:

> Now, Master Davy,
> Now, Sir, I have 'e;
> No one shall save 'e,
> Good master Davy.

Davy wrote casually in after-years, in a letter to his mother in which he was discussing a school for his young brother, John, that for himself he considered it fortunate that he was left much to himself as a child, put upon no particular plan of study, and enjoyed much idleness in Mr Coryton's school. 'I perhaps owe to these circumstances', he wrote, 'the little talents that I have and their peculiar application. What I am I have made myself.' He left in his will, to his old school at Penzance, a school to which the present Grammar School is heir, and which has recently taken his name, the annual interest on the sum of one hundred pounds. In a letter to his wife preserved in the Royal Institution he wrote that he wished such a sum to be divided every year amongst the boys of the Grammar School, on his birthday, which he hoped might be made a perpetual holiday for them. It was a good way of keeping his memory green. The boys of the Grammar School still enjoy their holiday. But while the pound has decreased in value, the number of boys in the school has enormously multiplied. Davy certainly intended a free day to each boy, and a gift of a little free private cash to each that each might enjoy himself after his own fashion. Lady Davy wrote that the gift was meant as an indulgence to the children. Perhaps he imagined a boy choosing to wander as he had wandered, with eyes and ears unconsciously active, and spirit open to the skyey influences of West Penwith. He had had in his own early days a favourite haunt, known by the name of Gulval Carn, a pile of rocks from which he had been able to see on one side the lines of the carns, and on the other the hedged fields, the Bay, and the Mount, ever changing with the changing light. It is a pity that the proposal made soon after Davy's death was not carried out – a proposal to perpetuate his associations with the Carn by making his memorial there. Valentine Le Grice, the same Le Grice who, with his brother Sam, had been school-fellow to Coleridge and Lamb at Christ's Hospital, composed in November

1831 verses to be inscribed on the rocks, but the work of setting the letters was never carried out. Now the rock is wound with ivy; violets and early potatoes tame the wildness of the approach.

For the last year of his school life – his school education ended when he was fifteen – Humphry was sent to the famous Grammar School at Truro, a sixteenth-century foundation, kept by the Reverend Cornelius Cardew. The present Cathedral School sprang from it. In the same letter quoted above, in which Davy speaks of the advantages of not having had too orderly a schooling under Mr Coryton, he recalls the joy with which he set out for Truro school, and the greater joy with which he left it. He excelled in Latin verse, and wrote sententious English essays after the manner of the time. He had, when he left school, enough knowledge of Greek for his imagination to be kindled, and a grounding in Latin. His mind would never be bumptious and provincial in relation to the past. But although he was not torpid at Truro, the essential processes of his mind were left untouched, he said, by the manner of his schooling. 'Learning', he wrote to his mother, 'is a true pleasure; how unfortunate then it is that in most schools it is made a pain. Yet Dr Cardew comparatively was a most excellent master. I wish John may have half as good a one. After all, the way we are taught Latin and Greek does not much influence the important structure of our minds.'

Dr Cardew is made vivid to us through such ardent old boys of the school as the Reverend Richard Polwhele, whose *Biographical Sketches in Cornwall* contains a biographical sketch of Davy, and is otherwise well worth reading, if only for its author's prejudices, which are on a heroic scale. He voices an early protest against the proposed widening of the curriculum in Grammar Schools, a widening which was being undertaken in the Truro school by Dr Cardew's successor. A more judicial estimate than Polwhele's of the Reverend Doctor may be read in *Cornelius Cardew, A Memoir*, by Sir Alexander Cardew (1926). Using his ancestor's diaries – 'May 23rd 1792. Threw off my wig' – and printing his detailed and curious Quinquennial Reviews of his state and affairs, Sir Alexander gives not only a portrait, pleasant and unpleasant, of a fighting character, at once pietistic and worldly, but also provides insights into the political and civic life of Truro during Davy's schooldays. In general he lights up the times. While still master of the Grammar School, and Rector of St Erme, Cardew acquired the living of Uny Lelant, putting in a Curate and taking a stipend that would have made any Jane Austenish young lady cast a speculative eye on him had he not been twice married already. He was a Whig who bitterly opposed the French war, and piously thanked God for an English victory at the close of it. He was twice Mayor of Truro, cared not a fig for unpopularity, was indefatigable in making interest for his family, and ruled his school with such severity that eleven boys ran away between 1773 and 1802. No boy ran away in Humphry's year. He was one of the twenty-five or so boarders in the Doctor's house at a time when the total strength of the school was about seventy. After Cardew's second term as Mayor of Truro the numbers began to fall off.

Humphry was remembered as a popular, good-tempered boy, ready to do other people's Latin exercises for them to save them from the Doctor's wrath. But he never ventured on drollery with Cardew as he had done with Coryton, nor could he have enjoyed much idleness under a man who, though he conscientiously produced scholars, confided to his diary his distaste for 'this irksome and laborious occupation in which, besides the anxiety arising from my wish to discharge my duty faithfully, I am exposed to the caprice and censure of many illiberal persons'. Mothers arrived to reproach him when he strove to discharge his duty too faithfully by their sons – '7th July, 1781, Mrs Osborne called about her son having been whipt'. It is said that Henry Martyn, later to become Senior Wrangler and famous missionary, was the only pupil of Cardew's who escaped being punished, when idle, with stripes on the back of his hand, and that was because Henry Martyn's hands were so thickly covered with warts that it was impossible to cane him. When Cardew was an old man he was still disputing; a year before Davy's death he was quarrelling with Mr Simmons of Lelant because Mr Simmons refused to let him take a tithe of the rakings of his barley. He seems to have been a just schoolmaster, for his boys bore him no ill-will in after life; but when he encouraged Humphry, on discovering that the boy wrote poetry, it must have seemed like being encouraged by a grampus. Whenever Davy grows piously sententious one feels that the Doctor is at his elbow.

Of people other than immediate relatives and schoolmasters who influenced Humphry, the most important was John Tonkin, who had been the adopted father of Grace Millett, and who became as it were the adopted grandfather of her eldest son. He was a Cornish moralist, with all that implies of the good and the tedious. While Humphry went to Mr Coryton's school he lived with Tonkin, and

Tonkin received in return vegetables and other fresh produce from Varfell, an arrangement made between close friends. In a memorandum book he jotted down expenses which he incurred for Humphry, and details of fruit, vegetables, and provisions he received from Varfell. Of this detailed account he wrote, 'This is not meant or intended as a charge against Mr Davy, only for my own satisfaction to know what Humphry has from me from time to time, and what I have from them.' It was he who paid for Humphry's schooling at Truro, the amount being £27. 12s. 10d.

The most vivid picture we have of Tonkin, who remained a bachelor, but who enjoyed the society of the ladies, is given by John Davy in *Fragmentary Remains of Sir Humphry Davy*, published in 1858, after the death of Lady Davy in 1855. John Davy thus describes his own earliest recollection of Tonkin – it must be remembered that when John was four, Humphry was sixteen:

> My earliest recollections carry me back to this respected friend, and he was a friend to us indeed. The memory of him is fixed in my mind not only because of his kindness to me as a child – he was fond of children and liked to have them about him – but also by his benign and venerable aspect, and his peculiar dress, that of the professional gentleman (by profession he was a surgeon), then passing away – the full wig, the sleeve and breast-ruffled shirt, the three-cornered hat, buckled shoe, etc. Well, too, do I remember the social meetings at his house, exclusively of the gentler sex, a certain number of whom regularly visited him, always drinking tea with him on a Sunday, all of whom he had known as children, and to whom and their children he was attached. Of these, two were sisters (Mrs Cornish and Miss Allen) both remarkable for their lady-like and pleasing manners, and one of them, unmarried, not less so for her perennial beauty. Such a little society could hardly fail to have an influence on a young mind . . .

Miss Allen was John Davy's godmother. The house in which these Sunday parties took place was the house in which Humphry Davy spent a good half of his time. A Mr Nicholls, a friend of the Davy family, wrote later, 'I have a vague recollection of Mr Tonkin in whose house Humphry seemed to be absolute, and who appeared very much attached to him.'

The house was a conspicuous one in Market Jew, and stood by the

Childhood and Schooldays

Old Market House, nearly opposite the Star Inn. It was not bequeathed to Humphry. John Tonkin left it to his great-nephew, Uriah Moore, who took the name of Tonkin in compliance with his uncle's will. Humphry Davy tried to buy it from Uriah Tonkin in later years; but J. A. Paris says that the interest which the Penzance Corporation possessed in the estate was an insurmountable obstacle. The 'Star' remains in Market Jew, although the portico on pillars, with the chamber overhead, a structure very like that at Little Keigwin, Mousehole, under which a man on horseback might find shelter, has been cut off. It was under this balcony at the 'Star' that Humphry Davy as a boy told stories to his friends, just as Scott told stories to his friends at the top of Arthur's Seat. Humphry's marvellous inventions were often founded on *The Arabian Nights*, but sometimes, too, no doubt, on the tales of mystery and horror then so much the fashion, and which John Tonkin's ladies could borrow from the Ladies' Book Club, established in 1770. There is more than a hint of Ann Radcliffe's style in Davy's descriptive prose. Of wonders and marvels there was also a mysterious and fearful local store. Some of Cornwall's best stories are told of the countryside Davy knew best – the story of Wild Harris of Gulval, for example. But Davy, like most people who tell stories aloud, was egged on to invent. He wrote later:

> After reading a few books, I was seized by the desire to narrate to gratify the passion of my youthful auditors. I gradually began to invent and form stories of my own. Perhaps this has produced all my originality. I never had a memory. I never loved to imitate but always to invent. This has been the case in all the sciences that I have studied. Hence many of my errors.

As he told his stories by the 'Star' he was very near a house supposed to be haunted. Running down towards St Mary's Chapel, to the left from Market Jew, was Chapel Street, once Our Lady Street. In this street a house was left uninhabited because of the dread of ghosts, and in this same street or near it little Maria Branwell, later to be the mother of the Brontës, was growing up. She was five years younger than Humphry Davy and was married in the same year; but there is no evidence that the two knew each other. John Davy says that when young people came to the haunted house in Chapel Street towards dusk they felt strange fears, and with beating hearts they quickened their footsteps.

[15]

Preaching quickened terror. I cannot find that either Davy's family or John Tonkin's was imbued with enthusiasm in the eighteenth-century religious sense of the word. The Branwells are named as members of the Methodist community, but not the Davys. In the person of the Unknown, in *Consolations in Travel*, Humphry seems to speak of himself as having 'been brought up in the ritual of the Church of England'. Mrs Davy and John Tonkin attended the old St Mary's Chapel with its spire whitewashed as a landmark. This chapel was on the site now occupied by the St Mary's dedicated in 1835, and which became the Parish Church of Penzance. As a very little boy Davy was taken to a meeting held between the regular church services by John Wesley when he came to the West to renew the tremendous influence he had gained among the Cornish, especially among the Cornish fishermen and miners. John Davy says it was handed down in the family that on this occasion Wesley, then a very old man, was attracted by the child and blessed him, placing his hands on his head and saying, 'God bless you, my dear boy'.

Another influence on Davy's early life was the coming of French refugees to Penzance. Rousseau and Voltaire died in the same year as Davy was born. He was nearly eleven at the time of the storming of the Bastille. His final year at school saw the execution of Louis XVI and Marie Antoinette; he was sixteen when, in 1794, Lavoisier was sent by the Revolutionary Tribunal to the guillotine. At that time Davy was being taught French by M. Dugast, an emigrant priest from La Vendée. Wordsworth reached early manhood when revolutionary hopes were bright; Davy when horror had replaced hope. His own eloquence was never to be denunciatory; he was to preach a faith insensibly forming in him during a childhood spent in a part of the country which, in a rural setting, was an advanced post of industrialism; a faith in the advancement of knowledge, and in the amelioration of hard labour by the use of machines.

Dr John Rowe, in his *Cornwall in the Age of the Industrial Revolution*, has emphasized the inventiveness of the Cornish practical engineers. In the intuitive nature of his genius Davy had something in common with the older and younger Trevithick; but it was the Scottish scientist and instrument-maker, James Watt — the associate of Dr Black, the visitor to Lavoisier's laboratory, the improver of the steam-engine by the invention of the separate condenser — who, through his son Gregory, helped to shape the direction of Davy's life. By the end of the year in which Davy was born, five of Watt's engines with separate condensers were working in Cornwall, and eight more were ordered. Watt himself came to Cornwall to supervise the erection of his engines and lived for a time at Plain-an-Guarry; William Murdoch was working at Redruth. When Davy took one of his many walks abroad, he might see not only the Mount but also the superstructure of the Wherry mine which had a shaft sunk below high-tide level. When he visited his relatives at St Just he was in the vicinity of the great Botallack mine, with already some submarine workings. Hayle Copperhouse was only about six miles from Davy's home. Speculation in tin and copper raised all kinds of hopes in families; a turn of Fortune's wheel might have made the Davys as wealthy as the Lemons, and Robert Davy to have had a street named after him. He 'ventured'; he was by nature a venturer, trying new ways of farming, and risking his money in tin. When he died in 1794 of a painful disease of the wind-pipe, he owed about £1,300.

Humphry at this time was nearly sixteen; there were four younger children. As was natural, Kitty was her elder brother's closest companion; Betsy was especially good to John, who was only four at the time of his father's death; Grace who was delicate, and who came in the middle of the family, might have felt a little isolated; but she did not. She said her mother's affection was equal for them all. Humphry thought all the children had ability.

Grace Davy was forty-four. She showed the courage and faith which helped to make it that both her sons regarded her always with peculiar tenderness. She left Varfell, set up a millinery business with a young refugee French lady in Penzance, took a lodger, and proceeded to bring up her family and pay off the debt.

Humphry, with the aid of John Tonkin, who paid the necessary sixty guineas 'consideration money', was apprenticed on February 10th, 1795, to John Bingham Borlase, surgeon, apothecary, and later physician of Penzance.

Although Humphry never liked the surgical part of his work, his apprenticeship was lucky. John Bingham Borlase happened to be a talented man, disinterested and zealous in his profession, and of standing among his neighbours. He was a great-nephew of the Rev. Dr William Borlase already noticed, and on the maternal side he was related to John Tonkin. When, in May 1813, he came to die, there was a notice of him in the *Gentleman's Magazine*. In this he was stated to have been an active promotor of the interests of Penzance. The enlargement of the quay was begun in his mayoralty; the dispensary was planned under his direction and conducted by him with great professional skill. His mind was well informed and his taste classically cultivated. He completed his medical studies and became a physician after Humphry had left Penzance. By becoming apprenticed to Borlase, Humphry continued to move in a circle of friends he knew well; but he had a chance of enlarging his knowledge and experience and was in communication with a formed and forward-looking mind.

In the Royal Institution many of his notebooks are treasured. They are notebooks covering different periods of his life, and no other memorials could bring a biographer more sensibly in contact with a mind. The very lack of method in the early books makes the contact closer. Now science, now poetry; now an elaborate essay, now a scheme for a story; now one side up of the notebook, now the other; now a current hand, now a hasty hieroglyphic or a drawing.

The scheme of self-education which he drew up is formidable; there is something touchingly youthful in the heroic reach of it. He had a natural zest for learning, but the circumstances of his family, his determination to help his mother, gave him a perseverance not commonly associated with an ardent temperament. His plan was set out in a little book; on one cover he drew a lyre; on the other a lamp wreathed with laurel or bay.

With Doctor Borlase: Penzance

In later years Davy was always to emphasize the interrelation of the different parts of the human mind, and the ties of connexion between the sciences. During his apprenticeship to Borlase poetry and metaphysical speculation mingled with his professional studies. Already it is clear from his earliest essays, and from significant entries in the notebooks, that he was prepared to overhaul and change his notions from time to time instead of allowing them to become beloved because they were his, or because he thought they were his, and so unalterable. If he was the child of a religious upbringing, home and school supporting one another, his youth belonged to the 'Enlightenment'. He read voraciously; he read the philosophers and argued with them; he wrote essays and argued with himself, scoring earlier paragraphs as he re-read them months later with, 'This is all frivolous'; 'This is false reasoning'. He argued with older, thoughtful men of the neighbourhood; readers of the Bible; men with an inward sense of the reality of religion; Quakers; or men educated through the admirable methodist class system. No doubt he sometimes heard that Cornish saying, 'He don't know enough to know that he don't know nothing.' In later years he was to write, 'The doctrine of the materialists was always, even in my youth, a cold, heavy, dull, and insupportable doctrine to me, and necessarily tending to atheism.' But he retained an eighteenth-century cast of mind; he was very much of the eighteenth century.

Humphry had the libraries of John Tonkin and his parson brother at his disposal, and he could borrow from the Book Club at Penzance; there was a Gentlemen's as well as a Ladies' Club by this time. Philosophy and mathematics were not neglected; but nor was poetry. I like the prospectus for a volume of poems which John Davy transcribed from a manuscript book of his brother's kept at this time.

Prospectus of a Volume of Poems

1st Eight Odes – 1. To the Memory 2. Sons of Genius 3. To St. Michael's Mount 4. Song of Pleasure 5. Song of Virtue 6. To Genius 7. and 8. Anomalous Yet

2nd Cornish Scenes – 1. St. Michael's Mount 2. Land's End 3. Calm 4. Storm

3rd A Tale – The Irish Lady

Although numbers seven and eight were 'anomalous yet' it was a volume of poems Davy was proposing, and a volume with some variety in it. At seventeen and eighteen he wrote poems in the manner

[19]

of Dryden, Gray, or Cowper; he used their patterns, but he described local scenes and attempted to tell a romantic story to which the name of a rock at Land's End was attached – the Irish Lady. He even amidst his 'awful shades' and 'rocks with verdure clad', brings in the homely tamarisk, and ventures a geological term – the *schistine* rocks. To criticize these poems, and to damn Davy as a poet because of them, is easy. Yet in themselves they compare favourably enough with the earliest verses of Wordsworth and Coleridge – though they also demonstrate the need for the revolution in poetic diction soon to be sponsored by Wordsworth. Davy's occasional poetry was always to lack the subtle fire which kindled his mind in the laboratory; he was not to become an inventor of harmonies. His poems are not discoveries. These youthful poems, ingenuous in their imitative art, are instructive to a biographer. In *The Sons of Genius*,[1] written at seventeen, can be heard through the stanza form of Gray's *Elegy*, that desire for glory which marked Davy's youth.

Soon after writing *The Sons of Genius* Humphry fell in love. J. A. Paris says it was with a young French lady living at Penzance – it will be remembered that Davy's mother had gone into business with a young French woman. In the notebooks appear, again and again, childishly drawn profiles of a face, a kind of doodling amidst the abstruse jottings. None of Davy's sonnets to his lady have been preserved, but in the poem *The Song of Pleasure* rosy Rapture and bright-eyed Pleasure usurp the place of the eagle. Ambition is dethroned by love. Davy first tunes the string of Dryden; but he continues in a strain which Keats – Cornish on his father's side – was to use:

> Let the philosophic sage,
> His silver tresses white with age,
> Amid the chilling midnight damp,
> Waste the solitary lamp,
> To scan the laws of Nature o'er,
> The paths of Science to explore;
> Curbed beneath his harsh control
> The blissful passions fly the soul.

[1] This poem, together with *The Song of Pleasure, Ode to St Michael's Mount*, and *Extract from an unfinished poem on Mounts Bay*, was printed in *The Annual Anthology*, vol. 1; ed. Robert Southey, Bristol, 1799.

With Doctor Borlase: Penzance

You, the gentler sons of Joy,
Softer studies shall employ!
He to curb the Passions tries . . .

The poem turns into an imaginative exercise what Humphry described to his brother John as a dangerous period of his life, a time when,
in John's words, 'he yielded to the allurements of occasional dissipation', the time between leaving Truro School and becoming an
apprentice. Humphry warned John of the danger, which seems to have
been the danger of diverting his energies from the pursuit of constant
knowledge to the pursuit of maidens with graceful ringlets, and
settling down to ordinary life. Humphry, like many elder brothers,
was inclined to be moral with his junior. Certainly his fancy was
haunted at this time by the nymph Theora. In his *Extract from an
Unfinished Poem on Mount's Bay* he transferred the maid and her blue
eye, which by now 'oft glisten'd with the tear of sensibility', to Erin,
and made her the heroine of the traditional story of the Irish Lady.
The Lady is said to have escaped from Ireland at the time of a massacre
of Irish Protestants by the Catholics in the reign of Charles II. She
was shipwrecked off Land's End near a rock, to which she clung after
all others in the ship had been drowned in the darkness. She was near
land but none could rescue her. At last she also died; but fishermen off
Land's End in stormy weather have seen her with a rose in her mouth,
sitting on the rock. Bernardin de St Pierre's *Paul et Virginie* was being
read and moving all hearts in Europe and England at about this time.
Davy wrote his own version of shipwrecked innocence and wild
scenery:

On the sea
The sunbeams tremble; and the purple light
Illumes the dark Bolerium, seat of storms.
High are his granite rocks. His frowning brow
Hangs o'er the smiling Ocean. In his caves
Th' Atlantic breezes murmur. In his caves,
Where sleep the haggard Spirits of the storm,
Wild dreary are the *schistine* rocks around
Encircled by the wave, where to the breeze
The haggard Cormorant shrieks. And far beyon
Where the great ocean mingles with the sky
Are seen the cloud-like Islands, grey in mists.

The Mercurial Chemist

> Thy awful height, Bolerium, is not loved
> By busy Man, and no one wanders there
> Save he who follows Nature, – he who seeks
> Amidst thy crags and storm-beat rocks to find
> The marks of changes teaching the great laws
> That raised the globe from chaos; . . .

There are other poems of this period. Davy liked to declaim. He did not write verse because he was miserable – the idea that a man must be in a mood of utter dejection and despair before he takes to poetry seems to have gone a good way towards killing it. There is no memory of him as a reluctant apprentice. On the contrary, he seems to have gone about his work with a ready cheerfulness, was sympathetic with his patients, was liked by the poor, by the old women, and even by the old men. Borlase's father used to cheer up at the sight of him, and sought his administrations. But sometimes he was lost to the muses. Borlase's sister told John Davy that once, when Humphry was on his way to visit a patient in the country, he flung away, in the exaltation of his chanted rhythms, the phial of medicine he had in his hand, and it was found next day in a hayfield near the path.

It was in the course of his profession that he came to the study of chemistry, and he came to it through books; but he came to this theoretic study after long years of almost unconscious, but exact and quick, observation of all the living forms around him, and of mineral forms too. In his poems he had expressed his feelings conventionally; but the feelings were none the less genuine and his senses acute – a good eye, delicate senses of taste and touch, but no ear; though later, when he was conducting exciting experiments, he would hum, not a tune, just an indeterminate humming.

Intellectually through the study of medicine, and practically through engineering, there was a link between Cornwall, Edinburgh, and Glasgow. An apothecary and surgeon in Penzance aspiring to become a physician would turn his thoughts towards Edinburgh. Because of his connexion with Borlase and the Tonkins, Humphry Davy in Penzance, separated from any centre of learning by a great length of what Josiah Wedgwood stigmatized as the worst road in England, began his education in the formal study of chemistry not with outmoded books such as fell to poor Jude's lot in Greek a century later, but with Lavoisier's *Traité Élementaire de Chimie*, published in France in 1789, and published in England in Robert Kerr's translation in 1790. Davy had sufficient French to read the book in the original; but he may have used the translation. Robert Kerr was of the same lineage as the Jane Kerr whom Davy was to marry.

Professor Douglas McKie's biography of Lavoisier, *Antoine Lavoisier: Scientist, Economist, Social Reformer* (1952), provides a narrative of the life of the great Frenchman who effected a revolution in chemistry; it also spreads out before us the particular enigmas which were occupying scientific minds at the time when Humphry was young, and the questions which were being posed in the effort to solve them. Humphry Davy was always to refer to the French chemist in later years as 'the great Lavoisier', and part of his prejudice against the French, a severity which distinguished him from the philosophical liberals with whom he was to be associated in Bristol, had one of its roots in the fact that Lavoisier was sent to the guillotine on May 8, 1794, by the Revolutionary Tribunal, and that certain of his old friends and associates 'showed themselves unwilling to take any action to protect and save him', though certain others in less high places 'did so in equal peril of their lives'. Dr McKie says that the *Traité* was a textbook for students, a summary of the new system of chemistry which Lavoisier had devised. The other book used by Davy was William Nicholson's *Dictionary of Chemistry* (1795). It had the advantage of giving Davy – always historically minded – some idea of the findings and theories of the past; findings and theories which were being displaced by the new ones, or were still ranging alongside the new.

Davy's mind was kindled. He was accustomed to wrestle with his authors and to argue with them on paper. Undeterred by Lavoisier's authoritative eminence, he began to argue with him as, in his reading, he had argued with the even more illustrious dead. Within a few months of beginning the theoretic study of chemistry, and before he had seen any experiment performed, he had imagined an hypothesis, plotted and attempted to carry out experiments of his own, and outlined the Essays which we shall notice when we come to consider his first published work.

His interest in the formal study of the sciences was immensely quickened by the coming to Penzance of Gregory Watt, James Watt's son by his second marriage. Already Penzance had a reputation for the mildness of its winter climate and Gregory Watt, because of his delicate health – he was consumptive – wintered at Penzance in 1797. He lodged at Mrs Davy's, and living as one of the family became Humphry's friend, supplying that intercourse with a trained contemporary mind – he was only the elder by a year or so – which Humphry, who did not go to a university, would otherwise have lacked.

Gregory Watt was just down from the University of Glasgow where he had read chemistry and geology. In West Penwith, so rich in material for a geologist, he and Humphry went on various expeditions to collect specimens of rock and minerals. In all kinds of ways up to the time of his early death in 1804, Gregory was a liberating influence on Humphry, who was to write of him later as the earliest and one of the dearest of his scientific friends 'for while I admired him as a philosopher I loved him as a man'. Gregory teased Humphry about his aspirations and quizzed his sententious moral pronouncements. Humphry needed to be teased, to be addressed, as Gregory addressed him with affectionate banter, as Dear H., My dear Philosopher; My dear Alchemist. On one occasion Gregory was Humphry's 'mystagogue in his initiation into the orgies of the mirth-inspiring Bacchus' – an initiation which nearly finished Humphry's devotions to Bacchus for good. On this occasion Humphry tried to recite his *Ode to St Michael's Mount* to Gregory.

For although Gregory was essentially a scientist and, not long before his death, produced a paper to be read before the Royal Society, he could also be absorbed and delighted by poetry without considering, as Humphry tended to do, whether or not it benefited his higher nature. With the copy of Burns which he sent to Davy he added a closer description of a reader's entrancement with poetry than has been common with some literary critics, who often seem to have betaken themselves to the criticism of poetry because they hate the stuff. He said he could begin, perhaps, with listless inanity, and then yield to the exquisitely insidious attacks of the beguiling enchantment. He describes how he finds himself unconsciously existing in a delightful maze through the imperceptible operation of charmed numbers. His death in 1804 at a time when Davy was lamenting to Coleridge the deleterious effect on himself of a too exclusive devotion to analysis was a misfortune. Davy, by nature sanguine, had not anticipated the death of his friend whose letters were full of spirit. To Clayfield, with whom he had worked in Bristol, Davy wrote, 'Poor Watt! He ought not to have died. I could not persuade myself that he could die; and until the very moment that I was assured of his fate, I could not believe that he was in danger ... He was a noble fellow, and would have been a great man. Oh! there was no reason for his dying – he ought not to have died.' In this same letter he set out the reasons for 'a kind of wild hope in the individual immortality of the better part of man', a hope to which he was to return again and again, and to elaborate finally in the essay entitled 'The Proteus', or 'Immortality' in *Consolations in Travel*.

While still at Penzance Humphry had also come to know Davies Giddy of Tredrea, St Erth, whose father had been curate of the Parish. Davies Giddy was an old boy of Penzance Grammar School, a graduate of Pembroke College, Oxford, a moderate Tory who had already been Sheriff of Cornwall, and who was to represent first Helston and then, for the greater part of his life, Bodmin in Parliament. At his marriage in 1817 with Mary Anne Gilbert he took the name and arms of his wife; so the name Gilbert swallowed up Giddy, and Gilbert he must be called. A pity; for his own name was in good Truro and Penzance repute. The Giddys, the Borlases, and the Tonkins must have been on familiar terms. Tredrea, still a handsome, well-placed house in St Erth parish, which is bordered by Ludgvan and St Hilary, is a delicious walk across country from Marazion, where lived Humphry's favourite aunt Millet and his sporting uncle Leonard. He and Leonard are said to have caught seven dozen trout in the little river and fishpond near Tredrea one June day. The number sounds like a fisherman's tale. John Davy, who tells it, was as mad about fishing as Humphry.

In addition to being Cornwall's representative and supporter, interested in her history and ancient manuscripts, Gilbert was a mathematician. He cared for all scientific and engineering ventures; he was the patron of Richard Trevithick, his theorist and paper-worker. He was also, at an important stage, of service to Davy. Their stories intertwine.

When he learnt that here was an experimenter, he invited Davy to his house, made him free of a library in which there were recent scientific and medical publications, and took him to Hayle Copperhouse where, in Mr Edwards' house, he saw for the first time a well-equipped laboratory including apparatus which he had previously seen only in engravings. It was here that Humphry, never one to hide his pleasure or his anger, expressed his 'tumultuous delight'; here that he worked an air pump 'with the simplicity and joy of a child engaged in the examination of a new and favourite toy'.

Either at Tredrea or among Borlase's books he came upon a publication jointly produced by James Watt and Dr Thomas Beddoes of Bristol: *Medical Cases and Speculations*, including Parts IV and V of *Considerations on the Medicinal Powers and the Production of Factitious Airs*, *1796*. In an Appendix was Samuel Latham Mitchill's *Remarks on the Gaseous Oxyd of Azote or of Nitrogene* (nitrous oxide) . . . in which Mitchill loosely tried to prove that nitrous oxide was the principle of contagion. Davy by 'a few coarse experiments' proved that this theory was incorrect. Again he wrote down his findings.

Both Gregory Watt and Davies Gilbert were friends of Dr Thomas Beddoes. Gregory was his patient, and Gilbert had been an undergraduate at Pembroke when Dr Beddoes, also of Pembroke, had been reader in Chemistry at Oxford. The two had become friends. While on a geological expedition in Cornwall Beddoes had visited Gilbert at Tredrea – in 1790 he had written his verses on a Cornish Lady who was prevailed upon to dance with a National Cockade. But he had not met Davy. It was through Gregory Watt that he came to read Davy's Essays, and to enter into correspondence with him.

Dr Beddoes thought he recognized in Davy's boyish experiments on heat and light, and in his experiments with nitrous oxide, signs of an uncommon talent; he was soon convinced that here was a young man who could be of service to, and could be served by, the Pneumatic Medical Institution which he was seeking, with the aid of various subscribers, to establish in Clifton. On July 11th, 1798, we find him writing to Davies Gilbert, 'I am glad that Mr Davy has impressed you as he has me. I have long wished to write to you about him, for I think I can open up a more fruitful field of investigation than anybody else. Should he not bring out a favourable result he may still exhibit talents for investigation, and entitle himself to public confidence more than by any other mode.' Beddoes sounded his subscribers, and wrote further letters to Gilbert and to Davy himself. The result was that in the fourth year of his apprenticeship, Davy was offered a position in the proposed Pneumatic Institution at Hotwells where, it was thought, he would be able to continue his medical education, and in connexion with which he would establish a laboratory to investigate the value of 'factitious airs' or gases in the treatment of disease. Some mild fun has been made of Davy because he asked for a 'genteel maintenance'; but the word 'genteel' had not become, at that time, through comic abuse, a vulgarism, though an advanced person like Beddoes might regard it quizzically. John Tonkin's lady friends would use it as naturally as Jane Austen, who was only three years older than Davy and was, at about that very time, engaged in writing *Northanger Abbey*. The name Hotwells is a reminder that in Jane's day and earlier – Gilbert White went to the Hot Wells to 'crisp his health' – Clifton was hoping to rival Bath as a spa, and that Blaize Castle was in the imagination of Catherine Morland an edifice as well equipped with towers, long galleries, and trap-doors as Udolpho. In *Emma* it was Bath *or* Clifton which Mrs Elton recommended to those out of health; and Maple Grove itself, though retired from the road, and with such an immense plantation all around it, was in the neighbourhood of Bristol. Upstarts from Birmingham, with fortunes made nobody knew how, were beginning to threaten the seclusion of Maple Grove: 'They came from Birmingham, which is not a place to promise much, you know, Mr Weston. One has not great hopes of Birmingham.' Throughout his life Davy was to use such words as elegance and decorum, respectable and genteel, with Miss Austen's own persistence. He had the sense of propriety which was part of the life of the age.

I wish there existed some record of the family discussions which must have taken place before Humphry set out for Bristol. John Tonkin certainly thought Dr Beddoes a subversive agitator and a quack. He could quite well have read a lampoon by Beddoes, who said the best way of getting his opinions accepted might be to manage his superiors by dedications and his inferiors by lampoons. He imagines just such an old-fashioned apothecary as Tonkin shouting at a patient who had taken a fancy to inhale the 'precious airs':

> You breathe indeed! do, d—mme, if you dare,
> Take one ounce measure of their cursed air!
> Look, here's your bill – pay that, Sir, – if you fail
> You take, by God, your next night's air in jail!
>
> *Clifton, Sept. 14, 1795*

Tonkin was old-fashioned, but Borlase was not. Although Davy had reached a point when an apprentice becomes useful, and begins to repay by his work some of the care bestowed upon him, Borlase did not seek to retain him. He surrendered Humphry's indenture, saying that he freely gave it up on account of the singularly promising talents which Mr Davy had displayed. He wrote of him that he was a youth of great promise, and said that he would not obstruct his present pursuits, which were likely to promote his fortune and his fame. The bond had been made between 'Grace Davy, widow of Robert Davy, Gent., of the parish of Ludgvan, John Bingham Borlase, and Humphry Davy'. Grace Davy, her son's legal guardian, gave her consent to break the agreement, and let Humphry go.

Davy set out for Clifton on October 2nd, 1798, going by coach through Bodmin and Launceston to Okehampton, where he breakfasted with Davies Gilbert. On the day of his departure Countess Spencer, wife of the first Lord of the Admiralty, was writing her galloping letter in praise of Nelson for his August 1st victory of the Nile: 'Joy, joy, joy to you, brave, gallant, immortalized Nelson . . . My heart is absolutely bursting with different sensations of joy, of gratitude, of pride, of every emotion that ever warmed the heart of a British woman, on hearing of her country's glory.' The Queen of Naples was writing that the world was mad with joy. At Okehampton Davy met a coach crowned with laurel, and streaming with ribbons, a coach coming down with victory, its trumpets triumphing in the great news. In Exeter the inhabitants were holding festival and the city was illuminated; whereas in Truro town, Davy's old schoolmaster, Cornelius Cardew, was having his windows broken by the mob for trying to prevent an illumination. They had a partial illumination in spite of him. Davy shared to the full his countrymen's enthusiasm for Nelson. Not only was he the great and victorious Admiral of the Fleet, but so keen a fly-fisher that after losing his right arm he learnt to angle with his left.

Like Nelson, Davy was a small man. As we shall gather from later observers as well as from John Davy, it was the play of expression in his face, its responsiveness to his thought and feeling, which made it attractive. John Davy says that when his brother's feelings were agreeable his face was 'eminently pleasing, I might say beautiful, for his smile was so, and his eyes were wonderfully bright, and seemed almost to emit a soft light when animated'. There are many other testimonies to Davy's eyes. Lockhart, not a romantic man, said they were the finest and brightest he had ever seen. Of his voice there is no such general praise; J. A. Paris says that at the time of his leaving Cornwall it was discordant. Paris also says he was round-shouldered. Both defects are likely enough in a fisherman-student who shouted out his verses against the wind and the sea.

Whatever the appearance of his shoulders, and the doubtful music of his Cornish accent in up-country ears, Davy went up to Clifton, a young man not quite twenty, ingenuous, confident, and aspiring; eager for glory, but with intent to serve mankind. With him he had his poems and his scientific essays. Paradoxically his politeness was in his poems and his impetuous enthusiasm in his essays.

3 · *With Doctor Beddoes: Clifton*

In Dr Beddoes Davy found a man more impetuous than himself. Coleridge was to note in 1803, 'Dr Beddoes' *first* Title Page for his work on the use of Cow-houses in consumption was, Speedy and certain Cure for Pulmonary Consumption this was altered during the printing, into Speedy Relief and probable Cure—lastly and so it stood, into Probable Relief and Possible Cure for pulmonary Consump'.[1]

The hurry and swing of Beddoes' mind were in piquant contrast with his person. Whereas his mind was ingenious and even sportive, his body was heavy and slow-moving. In a letter to his mother Davy set down his first impressions of Dr and Mrs Beddoes and of Clifton:

October 11th, 1798. Clifton.

My dear Mother,

I have now a little leisure time, and I am about to employ it in the pleasing occupation of communicating to you an account of all the *new* and *wonderful* events which have happened to me since my departure.

I suppose you received my letter, written in a great hurry last Sunday, informing you of my safe arrival and kind reception. I must now give you a more particular account of Clifton, the place of my residence, and of my new friends, Dr and Mrs Beddoes, and their family.

Clifton is situated on the top of a hill, commanding a view of Bristol and its neighbourhood, conveniently elevated above the dirt and noise of the city. Here are houses, rocks, woods, town and country in one small spot; and beneath us, the sweetly flowing Avon, so celebrated by the poets. Indeed there can hardly be a more beautiful spot: it almost rivals Penzance, and the beauties of Mount's Bay.

Our house is capacious and handsome; my rooms are very large, nice and convenient; and, above all, I have an excellent laboratory.

---

[1] *The Notebooks of Samuel Taylor Coleridge*, ed. Kathleen Coburn, vol. i, Note 1630.

With Doctor Beddoes: Clifton

Now for the inhabitants, and, first, Dr Beddoes, who, between you and me, is one of the most original men I ever saw – uncommonly short and fat, with little elegance of manners, and nothing characteristic *externally* of genius or science; extremely silent, and, in a few words, a very bad companion. His behaviour to me, however, has been particularly handsome. He has paid me the highest compliments on my discoveries, and has, in fact, become a convert to my theory which I little expected. He has given up to me the whole of the business of the Pneumatic Hospital, and has sent to the editor of the Monthly Magazine a letter, to be published in November, in which I have the honour to be mentioned in the highest terms. Mrs Beddoes is the reverse of Dr Beddoes – extremely cheerful, gay, and witty; she is one of the most pleasing women I ever met with. With a cultivated understanding, and an excellent heart, she combines an uncommon simplicity of manners. We are already very great friends. She has taken me to see all the fine scenery about Clifton; for the Doctor, from his occupations and his bulk, is unable to walk much. In the house are two sons, and a daughter of Mr Lambton, very fine children, from five to thirteen years of age.

I have visited Mr Hare, one of the principal subscribers to the Pneumatic Hospital, who treated me with great politeness. I am now very much engaged in considering the erection of the Pneumatic Hospital, and the mode of conducting it. I shall go down to Birmingham to see Mr Watt and Mr Keir in about a fortnight, where I shall probably remain a week or ten days; but before then you will again hear from me. We are just going to print at Cottle's in Bristol, so that my time will be much taken up, the ensuing fortnight, in preparations for the press. The theatre for lecturing is not yet open; but, if I can get a large room in Bristol, and subscribers, I intend to give a course of chemical lectures, as Dr Beddoes seems much to wish it.

My journey up was uncommonly pleasant; I had the good fortune to travel all the way with acquaintances. I came into Exeter in a most joyful time, the celebration of Nelson's victory. The town was beautifully illuminated, and the inhabitants loyal and happy. I was so pleased with Mr Russel and his family, and some other of the inhabitants to whom I was introduced, as to stay there two days, which I chiefly spent with Mr Russel. The morning after the illumination I rode round Exeter with Mr Russel, and was wonderfully delighted with the country, which is the most beautiful and highly cultivated of any I have yet seen.

It will give you pleasure when I inform you that all my expectations are answered, and that my situation is just what I could wish. But, for all this, I very often think of Penzance and my friends, with a wish to be there: however, that time will come. We are some time before we become accustomed to new modes of living and new acquaintances.

Believe me, your affectionate Son,

HUMPHRY DAVY.

The house was No. 3 Rodney Place, Clifton. Dr Beddoes was thirty-eight; his wife, Anna, was twenty-five. They had been married for over four years. Anna was a daughter of Richard Lovell Edgeworth of Edgeworthstown, Co. Longford, by his first wife, Anna Maria Ellers. She was a younger sister of Maria, whose *Castle Rackrent* was to be published in 1800. The *Parent's Assistant* had already appeared and was on the way to becoming the rage. Davy, Wordsworth, and Coleridge were all lucky enough to have preceded the first new era in education. All three recorded their joy, not in little moral tales made expressly for them but in the *Arabian Nights*, Coleridge stealing into the parlour at Ottery St Mary to take his copy from the window-seat as the sun looked in to make him feel safe from fear; Wordsworth saving up to try to get more volumes – he never saved enough; and Davy making up his marvellous additions to startle his friends.

The children referred to in Davy's letter were the children of Mr Lambton, a wealthy, philanthropic coalowner, who had been Dr Beddoes' patient and who had died in Italy. Together with Thomas Wedgwood he had been a principal subscriber to the proposed Pneumatic Institution. His two sons were living with Dr Beddoes for medical care and for education until they were old enough for Eton. Davy was used to children and liked their company; he talked with the Lambton boys as Penzance men had talked with him; both brothers said in later life how much they had unconsciously picked up from Davy's talk. John, the elder boy, was to become the first Earl of Durham and, true to the principles Beddoes had inculcated, the Radical Jack of the North; in 1839 he was the framer of the Durham Report recommending the grant of responsible government to the Canadians.

It was he who was the spokesman for the coalowners when a gift of plate was presented to Davy in recognition of his invention of the safety lamp, and of his refusal to take out a patent for it. John Lambton was the heir to vast wealth. Davy wrote on the leaf of a notebook kept at Clifton, in which are accounts of miscellaneous experiments, metaphysical speculations, and fragments of romances – 'Attentive was the youth, his eager eye darted bright lightning': 'I have neither riches nor power nor birth to recommend me, yet if I live I trust I will not be of less service to Mankind and to my friends than if I had been born with these advantages.' He fitted gradually into the Beddoes household. If he was surprised at Dr Beddoes, he was charmed with his wife. In a letter to his mother written after he had been longer at Clifton he said, 'You have been told he [Beddoes] is fond of money; I assure you it is quite the contrary, he is good, great and generous, and Mrs Beddoes is the best and most amiable woman in the world. I am quite naturalised into the family and I love them the more I know them.'

Perception of suffering had made of Dr Beddoes a man determined to find new ways of relieving pain and of removing ignorance, especially among 'the sick and drooping poor'. He was concerned both to combat the evils arising from drunkenness and the effects of hell-fire methods of conversion from it among hysterical patients. John Wesley himself left people feeling more alive than when he found them; but one has only to read the titles of tracts written by some of his so-called followers, or sing one of the more wormy hymns, to sympathize with Dr Beddoes' denunciations of converting by terror. He, a free-thinker, wrote a story to illustrate the misery caused by drunkenness, and the benefits of being sober. His *Isaac Jenkins, a Moral Tale*, which went into numerous editions, and which is in some ways mawkish, yet has some plain good pages which throw a vivid light on those conditions of the time which made radical reformers. It would be interesting to read the fragment by Beddoes with the good title, *Simple Simon and His Friend the Double-faced Cook.*

As a physician and disseminator of ideas Dr Beddoes had a considerable following among the young poets, the unorthodox in medical opinion, and extreme democrats in politics. It was his known radicalism – though the term philosophical radical did not come into use until later – at a time when to be even a moderate reformer was to be called a Jacobin, which made John Tonkin strongly oppose Davy's move to

Clifton. Beddoes was virulent in his political pamphlets. He had graduated at Oxford, had read medicine in Edinburgh, had taken his M.D. at Oxford, had been appointed to a readership in chemistry there, and had resigned partly because of his enthusiasm for the principles of the French Revolution. He had studied German and French, had travelled in France, had met Lavoisier, and had translated, among other things, the last paper Lavoisier wrote. He had a remarkable collection of books. Although Welsh by extraction, he had been born and brought up in Shropshire. In addition to his friendship with the Edgeworths he was associated with the progressives in Birmingham. A son of Joseph Priestley's was working with him. It was only about seven years earlier (July 14th, 1791) that Priestley's house in Birmingham had been wrecked by a Church and King mob, and Priestley had written to Sir Joseph Banks, President of the Royal Society, that, having lost his whole stock of chemicals, ores, minerals, etc., which he had for the purpose of his experiments, and desiring to replace them as quickly as possible, he requested Sir Joseph to bring his situation to the notice of his friends whose libraries were well supplied and to spare anything they could 'to set up a broken philosopher'. Sir Joseph does not seem to have done anything to help the broken philosopher who had to seek refuge in America. Nor would Sir Joseph lend his name or give his money to furthering Dr Beddoes' Pneumatic Medical Institution. In 1794 he had been invited to do so both by the Duchess of Devonshire, who was much impressed by Beddoes' experiments and thought they should be given a fair and open trial, and by James Watt. To the Duchess Sir Joseph regretted that he could not support causes of which he did not approve; apart from the Doctor's revolutionary principles he considered the proposed treatment likely to do more harm than good. When James Watt solicited his patronage for the Institution he said, in his refusal, that he was always careful not to annex his name to any scheme unless convinced of the probability of its success; he thought better proofs should be laid before the public before they were asked to divert their charity. Sir Joseph was an experimentalist in agriculture; but as a practical farmer on his great estates he understood the slow operations of time. As Lavinia, Countess Spencer wrote to him in October 1800, 'how wonderfully slow John Bull is in altering his old ways! So much the better. I am sick to death of that incessant love of change which seems to possess every nation under heaven except this wise and cautious island.' Beddoes thought this caution in the medical profession had resulted in ignorance that would disgrace a savage.

One of Beddoes' most interesting older friends was Dr Erasmus Darwin of Litchfield, Charles Darwin's grandfather. Davy had referred to Erasmus Darwin's *Zoonomia* in his Essays. To Erasmus Darwin Beddoes soon wrote of Davy, 'I think him the most extraordinary person I have seen for compass, originality and quickness of thought.' This was no mean praise. Beddoes had had unusually wide experience of outstanding young men. He hurried on the publication of Davy's Essays. Davy did not reach Clifton until the second week in October; in January of the new year Beddoes published – Joseph Cottle was the printer – a volume of miscellaneous essays which he entitled, *Contributions to Physical and Medical Knowledge, principally from the West of England.* In this volume he gave pride of space to his young superintendent's researches – 'infant speculations' Davy was to call them later. They are reprinted in the second volume of Davy's *Works*, and they consist of two or, in a way, three related essays: *Essay on Heat, Light, and the Combinations of Light with a New Theory of Respiration* and *An Essay on the Generation of Phosoxygen, (oxygen gas) And on the Causes of Colours in Organic Beings.* It must be remembered that Davy came to the study of physics and chemistry by way of physiology and medicine, and was at least as much interested in respiration as in combustion; while light, which has a peculiar quality of radiance in West Penwith, as though it were light of light, very light, had held his imagination from childhood. Light and time are almost Davy. In a Clifton notebook he wrote, 'What we mean by Nature is a series of images: but these are constituted by light. Hence the worshipper of Nature is a worshipper of light.' Lavoisier, he thought, had insufficiently considered the chemical effects of light, chemical effects not less important than the physical. He began with experiments on the generation of heat, seeking to answer Lavoisier's query, 'La lumière, est elle une modification du calorique, ou bien, le calorique, est il une modification de la lumière?' As early as 1774, four years before Davy was born, Joseph Priestley had isolated the gas which, in 1779, Lavoisier named oxygen, after having demonstrated that it supported flame and respiration, and formed one-fifth of the common air. Priestley, holding it to be common air deprived of phlogiston, had called it dephlogisticated air, that is, air deprived of the peculiar principle of inflammability. To the end of his life Priestley remained a phlogistinian. The existence of phlogiston, a hypothetical substance, the imaginary stuff of flame, was so much taken for granted in eighteenth-century science that the word phlogiston could be readily used as a metaphor – Coleridge accused Southey of having phlogiston in his heart when his fellow-poet was in a flaming rage. Lavoisier denied in his theory of combustion the existence of phlogiston, but gave, in his *Table of Simple Substances belonging to all the Kingdoms of Nature which may be considered as the Elements of Bodies*, one to which he assigned the name *calorique* – in English, caloric. The youthful Davy discarded phlogiston and caloric or any stuff of flame, any matter or material cause of the sensation of heat. For the sensation of heat he kept the common name heat; for the cause of the sensation, going back a century to Robert Boyle, he tried to show that it was not a mode of matter but a mode of motion – he named it repulsive motion. Thus in his first essay he joined those of his contemporaries who, dissatisfied with the evidence produced in favour of the existence of an igneous fluid, and perceiving the generation of heat by friction and percussion, supposed it to be motion. Davy set out details of experiments which demonstrate his resourcefulness. Then, having proved as he thought the non-existence of caloric among the subtile ethereal fluids, he turned to consider light as a substance acting the most important part in the economy of the universe and the life of man. He supposed light to combine with oxygen, and he supposed the electric fluid to be condensed light. Oxygen combined with light he named Phosoxygen. He propounded a theory of respiration and, in his second essay, related a series of experiments made in Penzance on marine plants, having been led by analogy to infer that they perform in the sea the same part for aquatic animals that terrestrial plants had previously been supposed to do for land animals – the important part of renovating the oxygen consumed in the function of respiration. John Davy believed that the confirmation of this analogy was the first discovery his brother made. In Addenda to his Essays Davy remarked that his experiments on the generation of heat were made before the publication of Count Rumford's paper on the heat produced by friction – a statement endorsed by Beddoes. Count Rumford's experiments alone, Davy wrote, went far to prove the non-existence of caloric and, when compared with the second and third experiments in his own Essay, would leave no doubt, he conceived, on the minds of the impartial and philosophic reasoner.

With Doctor Beddoes: Clifton

Rumford was a scientist with an international reputation and a fellow of the Royal Society; Davy was a country youth at the time of these experiments. I am told that Rumford's well-known and less-well-known experiments in this field remain among the most interesting of their kind ever devised and carried to a successful conclusion; and that in comparison with this master's work Davy's 'infant speculations' were what Davy himself justly called them. It would be unnecessary to dwell on these early Essays were it not that three of the experiments described, the first devised to show that light is not an effect of heat, the second and third the celebrated ice-melting experiments, are still commonly cited in textbooks as being of major importance in constituting experimental proof of the dynamical theory of heat, although in a historical note, 'Humphry Davy's Experiments on the Frictional Development of Heat' (*Nature*, March 9th, 1935, p. 359), Professor E. N. da C. Andrade demonstrated in an incisive way that they should cease to be so considered. He points out that it in no way detracts from the subsequent greatness of Davy to show that his first experiments, 'carried out when he was a country lad, were uncritical and lacked all quantitative basis'. 'They should', he says, 'cease to be ranked with such convincing demonstrations as those of Rumford.' Dr Douglas McKie (*Nature*, May 25th, 1935, p. 878) makes further observations on persistent textbook errors concerning Davy's earliest experiments and on his apparatus and instruments as described by his earliest biographers. Davy did not melt ice by friction in a vacuum, though he did make some kind of boyish attempt at doing so.

The immediate reception of Davy's Essays was like a defeat in battle. Davy was laughed at. He was laughed at partly with some reason, and partly because there were merry ruffians in the scientific as in the literary world who were ready to barb their shafts at a satellite of Beddoes. It did not mollify Davy that some of the criticism was ignorant. He would have agreed with Dr Johnson:

> Fate never wounds more deep the generous heart,
> Than when a blockhead's insults point the dart.

In his notebook he drafted excuses for himself and accusations of his critics. He was young and angry; his punctuation foundered and some words are undecipherable:

> These critics perhaps do not understand that these experiments were made at a time when I had studied Chemistry but for four

[37]

months when I have never seen a single experiment executed and when all my chemical information was derived from Nicholson's Chemistry and Lavoisier's Elements. They do not perhaps consider that my apparatus could not be made more perfect and that infinite labour was required in performing every experiment
They have   that their inaccuracy have been determined by
        eminent Chemists, who are these eminent chemists, there are few enough God knows in England. They were often repeated and they will in general be found to be accurate. In a system such criticism I despise. I should have passed it unnoticed had I not considered the influence of it on society, had it not been a lamentable consideration to me that so little was the candour of an English reviewer   The simple labour of a young man of nineteen without . . . had they come through the medium of the Royal Society. . . . I was perhaps wrong in publishing with such haste a new Theory of Chemistry and my mind was ardent and enthusiastic, I believed that I had discovered the truth, since that time my knowledge of facts is increased, since that time I have become more sceptical   Yet on reviewing my theory I find reasons for relinquishing but a very small part of it.

Prejudice and ignorance her eternal concomitant may for a while arrest the progress of the human mind. The criticism of [closet?] chemists may influence the world, men who never made an experiment themselves may sit in judgment on the labour of the man who devotes his life to the interest of science and of mankind may be neglected and persecuted. It is not in the power of error to destroy truth. I would always wish for existence of a certain portion of scepticism.

Davy did not publish his anger or his self-justifications. He had met, he said, friends who cheered him:

I have felt the warmth,
The gentle influence of congenial souls
Whose kindred hopes have cheer'd me; who have taught
My irritable spirit how to bear
Injustice. . . .

Soon he was writing that when a man is ardent he is not the best judge of the effort he makes. And if Davy was ridiculed, he was also praised, and by one whose praise was worth having. Joseph Priestley, from

With Doctor Beddoes: Clifton

America, wrote in an Appendix to his chemical work *The Doctrine of Phlogiston Established*, 'When some progress was made with the printing of this work, I met with Dr Beddoes' *Contributions to Physical Knowledge*, and in it Mr H. Davy's *Essays* which have impressed me with a high opinion of his philosophical acumen. His ideas were to me new, and very striking, but they are of too great consequence to be decided upon hastily.' This should have been balm, one would have thought, for any mortification. But Davy, throughout his life, without possessing exactly a self-wounding imagination, was extremely sensitive to censure, or, perhaps, to the fact of having laid himself open to it. No one else regretted his early Essays as permanently as he did himself. He told Dr Hope that he would joyfully relinquish any little glory or reputation which he might have acquired by his subsequent researches in Clifton were it possible to withdraw these first essays and remove the impression which he fancied then they were likely to produce. They mortified him living and they have dogged him dead. Ironically, he has through all the years almost till now been rather in the same position as a grave poet embarrassed by a constantly anthologized example of his lyrical sweet youth.

In a letter to Dr Nicholson published in *Nicholson's Journal*, February 1800, Davy begged to be considered, until he had satisfactorily explained certain facts by new experiments, a sceptic with regard to his own particular theory of the combinations of light, and theories of light in general, and he ceased to make use of the nomenclature he had suggested. 'Consistency in opinion', he was to write, 'is the slow poison of intellectual life, the destroyer of its vividness and energy.'

Elasticity of mind and mood were of the very essence of Davy's being in his early days at Clifton. At the same time as he was preparing his Essays for the press he was entering enthusiastically upon the researches which had brought him to Clifton as the associate of Dr Beddoes. Priestley had directed the forces of his mind to pneumatic chemistry; Beddoes wished to harness pneumatic chemistry to medicine. Hence the founding of the Pneumatic Institution, the design being to establish a laboratory for research, a theatre for lecturing, and a hospital for patients. The whole scheme was experimental. Thomas Wedgwood, who had come to Somerset to be near Beddoes, was so anxious that the inhaling of gases in the treatment of disease should be regularly tried and tested that he considered the money would be well spent even if the results obtained were negative, or if it were proved that factitious airs could not have medicinal value. He himself was painfully and personally interested, for he suffered from an incurable disease and hoped to gain relief from the new theories of respiration. For a time he was advised by Beddoes to lodge over a butcher's shop, as a healthy situation.

James Watt, like Thomas Wedgwood, had a dear incentive to quicken his interest in the possibilities of gases in the treatment of disease. His son Gregory might benefit; he might be cured. Watt had devised an apparatus for the manufacture of the airs and, in 1796, had issued a pamphlet for their use. A letter from Watt to Davy at Clifton gives a minute description of a cabinet for breathing oxygen and other 'precious' airs. But before this Davy had received an invitation to visit James Watt. At So-Ho in Handsworth, and in Birmingham, he had met engineers and scientists of eminence whom, in his letters home, and to Davies Gilbert, he excitedly praised. This meeting with Watt, Keir and others in a city where applied science was being so eagerly pursued must have been of service to him apart from the fitting up of his laboratory at Clifton. It was in Leeds that Dr Priestley had been led, through watching the process of fermentation in a brewery next door to his house, to give his leisure to the study of pneumatic chemistry. The Greek word *pneuma*, wind, breath, spirit, which has become, through its dreary compounds, so commonplace for us, was a vital word all new and fresh to the chemists of the Romantic period. Davy, living by light and drawing breath, was always conscious of the mystery. If one could win to such an understanding of vital processes as to be able to prescribe in cases of disorder the inhaling of the right remedial factitious air, medicine might be rescued from empirical practice to become an exact science. 'Thus would chemistry', he had written at the close of his first Essay, 'in connection with the laws of life, become the most sublime and important of all sciences.'

Before he could even find a house in which to carry out his researches Davy had to overcome ignorances and prejudices; but by the beginning of 1799 a house had been found in Dowery Square in the then fashionable quarter of Hotwells. It was fitted up for the purposes of the Institution and provided with a laboratory. Here Davy carried out a series of investigations with great thoroughness. It became his duty, as superintendent of the Medical Pneumatic Institution, to investigate the physiological effects of gases and to consider their chances as useful agents in medicine. He began his researches in April 1799; he worked at them incessantly for ten months, and he spent three months detailing them. The work was published in a separate volume in 1800, and was entitled: *Researches, Chemical and Philosophical; Chiefly Concerning Nitrous Oxide, or Dephlogisticated Nitrous Air, And Its Respiration.* Davy's own *Introduction* to his *Researches* was dated June 25, 1800, from his house in Dowery Square, Hotwells, Bristol. The Fourth Research, in which the effects of breathing artificial airs, and especially the pleasure-producing air, as he named nitrous oxide, are described, is still interesting to read because of the intensity of the search for words to describe sensations so novel that Davy wrote in a notebook – 'like blind men who use the language of sight'. After relating in exact detail the conditions of an experiment he writes an account of the effects:

. . . A thrilling, extending from the chest to the extremities, was almost immediately produced. I felt a sense of tangible extension highly pleasurable in every limb; my visual impressions were dazzling, and apparently magnified, I heard distinctly every sound in the room, and was perfectly aware of my situation. By degrees, as the pleasurable sensation increased, I lost all connection with external things; trains of vivid visible images rapidly passed through my mind, and were connected with words in such a manner as to produce perceptions perfectly novel. I existed in a world of newly connected and newly modified ideas. I theorised; I imagined that I made discoveries. When I was awakened from this semi-delirious trance by Dr Kinglake, who took the bag from my mouth, indignation and pride were the first feelings produced by the sight of the persons about me. My emotions were enthusiastic and sublime; and for a minute I walked about the room perfectly regardless of what was said to me. As I recovered my former state of mind, I felt an inclination to communicate the discoveries I had made during the experiment. I endeavoured to recall the ideas, they were feeble and indistinct; one collection of terms, however, presented itself: and with the most intense belief and prophetic manner, I exclaimed to Dr Kinglake, 'Nothing exists but thoughts! The universe is composed of impressions, ideas, pleasures and pains.'

When he had respired the gas after being fatigued by a journey, he had had, as he began to lose consciousness, such a vivid and intense recollection of some former experiments passing through his mind that he had called out, 'What an amazing concatenation of ideas!' Davy was, throughout his life, interested in dreams and visions. He often felt great pleasure, he said, when breathing the gas alone, in darkness and silence, occupied only by ideal existence; and when he breathed the gas after excitement from moral or physical causes, the delight he felt was often intense and sublime. He describes how one May night, after walking for an hour amidst the scenery of the Avon, at this period rendered exquisitely beautiful by bright moonlight, his mind being in a state of agreeable feeling, he respired six quarts of newly prepared nitrous oxide:

> The thrilling was very rapidly produced. The objects round me were perfectly distinct, and the light of the candle not as usual dazzling. The pleasurable sensation was at first local, and perceived in the lips and about the cheeks. It gradually, however, diffused itself over the whole body, and in the middle of the experiment was for a moment so intense and pure as to absorb existence. At this moment, and not before, I lost consciousness; it was, however, quickly restored, and I endeavoured to make a by-stander acquainted with the pleasure I experienced by laughing and stamping. I had no vivid ideas. The thrilling and the pleasurable feeling continued for many minutes; I felt two hours afterwards, a slight recurrence of them, in the intermediate state between sleeping and waking; and I had during the whole night vivid and agreeable dreams. I awoke in the morning with a feeling of restless energy, or that desire of action connected with no immediate object, which I had often experienced in the course of experiments in 1799.

He was concerned with the difficulty of finding words to describe the strange effects of inhaling nitrous oxide. He realized that, from the nature of the language of feeling, the details he gave of his experiences contained many imperfections. These induced pleasures or pains could only be intelligibly detailed when associated, during their existence, with terms standing for analogous feelings. He had, he said, sometimes experienced from nitrous oxide sensations similar to no others and consequently indescribable. Other persons expressed the same difficulty in finding words. In the hospital he and Dr Beddoes had tried the gas on paralytic patients. One of these, when asked what he felt after breathing nitrous oxide said, 'I do not know, but very queer'; a second patient said, 'I felt like the sound of a harp.' With this last remark may be compared the evidence of Dr Beddoes who shouted out 'Tones!' on the only occasion when, by breathing several doses of nitrous oxide in quick succession, he came near unconsciousness. Davy included in his book detailed descriptions furnished by various people of what they felt like while under the influence of the gas. Many of these people were his friends. But before turning to them, and to the manifold other interests of Davy in Clifton, here is a quotation from William James's *The Varieties of Religious Experiments*, showing how a psychologist, a century later than Davy, described his experience after breathing nitrous oxide:

> Looking back on my own experiences [under the influence of nitrous oxide], they all converge towards a kind of insight to which I cannot help ascribing some metaphysical significance. The keynote of it is invariably a reconciliation. It is as if the opposites of the world, whose contradictoriness and conflict make all our difficulties and troubles, were melted into unity. Not only do they, as contrasted species, belong to one and the same genus, but *one of the species*, the nobler and the better one, is itself the genus, and so soaks up and absorbs its opposite into itself.

A harmony indeed! No wonder Beddoes exclaimed, 'Tones!' A Mr Wansey was much more definite in using a musical analogy to convey his feelings. He said that after experiencing fullness in the head he went on to sensations so delightful that he could only compare them to those he had felt (being a lover of music) about five years before in Westminster Abbey, in some of the choruses of *The Messiah*, from the united powers of 700 instruments. Other breathers used other comparisons. Tom Poole compared his pleasant feelings and increased powers of body and mind to those he had experienced in less degree when ascending some high mountains in Glamorganshire. S. T. Coleridge felt a highly pleasurable sensation of warmth stealing over his whole frame, resembling that which he remembered once to have experienced after returning from a walk in the snow into a warm room. Another time his heart beat as if it were leaping up and down. To discover what effect breathing the gas had on his impressions he (Coleridge) said he fixed his eyes on some trees in the distance, but did not find any other effect except that they became dimmer and dimmer, and looked at last as if he had seen them through tears. It is amazing how reckless they all were of their bodies. Coleridge once inhaled after a hearty dinner. Davy tried it after drinking a bottle of wine in large draughts in less than eight minutes, the effect being that in less than an hour he sank into a state of insensibility and remained thus for two hours or so. He observes in a footnote that the powerful effect of a mere one bottle was due to the fact that he was accustomed to drinking water, and had never been but once completely intoxicated before in the course of his life. That was the celebrated occasion when he was initiated by Gregory Watt. He found on the whole that breathing the gas was good for a hang-over; at least it did not increase his debility in the way debility from intoxication by two bottles of wine is increased by a third. Often it seems to us, as we read, as though the consequence of breathing the gas was merely an intensification of the natural characteristics of the breather. Even the profession of the patient may be guessed at. James Webbe Tobin, brother of John Tobin, the dramatist, found that his spirits continued much elevated for several hours after one of his doses. His step was firm, and all his muscular powers increased. His senses were more alive to every surrounding impression; he threw himself into several theatrical attitudes, and traversed the laboratory with a quick step; his mind was elevated to a most sublime height. He wrote, 'It is giving but a faint idea of the feelings to say that they resembled those produced by a representation of an heroic scene on the stage, or by reading a sublime passage in poetry when circumstances contribute to awaken the finest sympathies of the soul.' Quick sensibility played a part. Coleridge on one of his occasions found more unmixed pleasure than he had ever before experienced; Boulton and Watt were hardly moved at all.

Fairly general in all breathers were involuntary laughter and the antic movements which have given nitrous oxide the modern name of laughing gas. Davy found it better to desist with females inclining to hysteria and with those of too delicate sensibility. The falsifications of active fancy had to be reckoned with, and self-induced feelings.

But then how hard it is for man, or woman either, to be even moderately truthful! We find young Citizen Southey, for example, writing to his brother Tom of the effects of the gas in the most uninhibited tone. The 'rigidly virtuous' young republican wrote, after imbibing, 'Oh, Tom! such a gas has Davy discovered, the gaseous oxide. Oh, Tom! I have had some; it made me laugh and tingle in every toe and finger-tip. Davy has actually invented a new pleasure, for which language has no name. Oh, Tom! I am going for more this evening! It makes one strong and so happy! So gloriously happy!' His evidence, printed in Davy's book, is quite different. He confesses to a feeling of apprehension of which he is unable to divest himself; and although his involuntary laughter was pleasant at first and, during the remainder of the day, he imagined that his taste and hearing were uncommonly quick, and he felt more than usually strong and cheerful, yet when he tried again, after an interval of some months, during which his health had been impaired, the effect of the gas on him was adverse. He felt that the quantity which he had formerly breathed would now have destroyed him, and he concluded his account with the words, 'The sensation is not painful, neither is it in the slightest degree pleasurable.'

Many of the accounts are interesting to read today because of what we know of the subsequent careers of their authors. Dr Peter Mark Roget, although his *Thesaurus of English Words and Phrases* was not published until 1852, had projected his system of verbal classification fifty years earlier. He worked with Dr Beddoes and corresponded with Davy, to whom he described in a letter the effect on himself of inhaling nitrous oxide. It is not, I think, fanciful to see in this letter something of the care-to-be-exact characteristic of the man who was to classify English words and idioms according to the ideas they expressed.

The effect of the inspirations of nitrous oxide was that of making me vertiginous, and producing a tingling sensation in my hands and feet: as these feelings increased, I seemed to lose the sense of my own weight, and imagined I was sinking into the ground. I then felt a drowsiness gradually steal upon me, and a disinclination to motion; even the actions of inspiring and expiring were not performed without effort: and it also required some effort of mind to keep my nostrils closed with my fingers. I was gradually roused from this torpor by a kind of delirium, which came on so rapidly that the air-bag dropped from my hands. This sensation increased for about a minute after I had ceased to breathe, to a much greater degree than before, and I suddenly lost sight of all the objects round me, they being apparently obscured by clouds, in which were many luminous points, similar to what is often experienced on rising suddenly and stretching out the arms, after sitting long in one position . . .

What Dr Roget expressed when he said he seemed to 'lose the sense of his own weight', Anna Beddoes described when she said that after inhaling she frequently felt as though she were ascending in a balloon, and found she could walk more easily up Clifton Hill.

Davy's conclusions on the use of inhaling nitrous oxide are temperately stated. He writes, 'As nitrous oxide in its extensive operation appears capable of destroying physical pain, it may probably be used with advantage during surgical operations in which no great effusion of blood takes place.' In his book *The English Pioneers of Anaesthesia* Dr F. F. Cartwright, commenting on this sentence, and on words which he quotes from a notebook of Davy's – 'removing physical pain from operations' – has emphasized that they constitute the first recorded suggestion of a practical means of anaesthesia. He traces in detail the work of Beddoes and of Davy in this field, and relates it to subsequent discovery and practice.

But a practical means of anaesthesia was not, of course, what Beddoes and Davy were consciously seeking. Beddoes wanted exact and tested knowledge of how gases affected health and spirits; but he was not himself an exact man. He was not, Davy said, much given to experiment and little attentive to it. Beddoes was promoter and seer. He encouraged methods of discovery so that he might make use of new and incontrovertible facts about organic life without which, he said, 'our notions of the living world will, in my opinion, continue to be as confused as the elements are said to have been in chaos.' 'On some future occasion', he wrote, 'I may presume to point out the region through which I imagine the path to wind, that will lead the observers of some future generation to a point whence they may enjoy a view of the subtle, busy and intricate movements of the organic creation as clear as Newton obtained of the heavenly bodies'.

Davy's book, in which this observation by Beddoes appears in a footnote, was published in the summer of 1800; it forms volume iii of his collected *Works*. In the concluding paragraph of his Introduction he expressed his obligation to Dr Beddoes: 'I cannot close this Introduction without acknowledging my obligations to Dr Beddoes. In the conception of many of the following experiments, I have been aided by his conversation and advice. They were executed in an Institution which owes its existence to his benevolent and philosophic exertions.'

4 · Clifton Company

The now very widespread use of nitrous oxide as an anaesthetic in dental operations – Davy tried it when he had toothache – does not seem to excite all the remarkable effects recorded in Davy's time. The explanation may be that frequently the gas was diluted with common air, and administered in small doses, so that only the excitable stage preceding unconsciousness was reached. Perhaps, too, the powers of suggestion were at work. The investigation was, in its very nature, sensational. Davy's own records – in a notebook of the period – especially when he is trying to distinguish between the agency of nitrous oxide and that of 'other physical and moral causes', have a heady note subdued in the published accounts. The entry for April 27th reads: 'This evening April 27th I have felt a more high degree of pleasure from breathing nitrous oxide than I ever experienced from any cause whatever – a thrilling all over me most exquisitely pleasurable, I said to myself I was born to benefit the world by my great talents . . .' The entry for May 5th is, 'May 5th: After eating a supper, drinking two glasses of brandy and water, and sitting for some time on the top of a wall by moonlight reading Condorcet's *Life of Voltaire*, I requested Mr Dewyer to give me a dose of air . . .'

According to Joseph Cottle, Davy seemed at this time, when experimenting on himself, to act as if, in sacrificing one life, he had two or three others in reserve on which he could fall back in case of necessity. 'A more persevering and enthusiastic experimentalist than Mr Davy', he writes, 'the whole kingdom could not have produced; an admission which was made by all who knew him, before the profounder parts of his character had been developed. No personal danger restrained him from determining facts, as the data of his reasoning; and if Fluxions, or some other means, had not conveyed the information, such was his enthusiasm, he would almost have sprung from the perpendicular brow of St Vincent to determine his precise time in descending from the top to the bottom.' Cottle thought the experiments Davy engaged in during his investigation of gases produced the affections of the chest to which he was subject through life and which, Cottle thought, beyond doubt shortened his days.

The joyous manifestations at the Pneumatic Institute gave rise to a *jeu d'esprit* entitled *The Pneumatic Revellers: An Eclogue*.[1] The scene is the Medical Pneumatic Chambers, and among the characters are Dr Beddoes, the Rev. R. Barbauld, Mrs Barbauld, the 'children's friend' so much disliked by Charles Lamb, and Robert Southey. Mrs Barbauld is made to drink of the gas and to exclaim:

> . . . Blithe as when I skipped with Lissy
> Crowned with many a pretty flower,
> Beddoes! how I long to kiss y'
> In my trembling moonlight bower.

Southey in an even riper stage of delight, ineffably moved, is provided with even more deliciously absurd words:

> . . . I spurn, I spurn
> This cumbrous clod of earth; and, borne on wings
> Of lady-birds, all spirit, I ascend
> Into the immeasurable space.

Most of the Beddoes' friends tried the gas, and these friends, as the detailing of the experiments prove, were of the greatest diversity of temper and talent. Among them was Thomas Wedgwood who, with his brother Josiah, frequently visited another brother at Upcott. They were the sons of Josiah Wedgwood, the potter of Etruria. Davy was at this time too full of health and energy – his titanic energy reminds one sometimes of the later Dickens – to be fully sympathetic always towards Thomas Wedgwood. He was sometimes harsh towards his settled sadness. In Bristol, Davy's were the easy virtues springing from a genial feeling of health; whereas Thomas Wedgwood's virtues, Coleridge truly said, and Coleridge of all men had reason to know, were exercised in the barrenness and desolation of his animal being. But it must have been immensely beneficial to Davy to know Wedgwood; he did not exaggerate when he said that Wedgwood's advice was a secret treasure to him. Nor was he indebted merely for advice – although as one reads of this group of Englishmen one concludes that man is indeed 'an advising animal'. Just to know a man who combined

---

[1] *Biographical Sketches in Cornwall*, R. Polwhele, 1831.

so equally in his nature what Coleridge noted as a fine and ever-wakeful sense of beauty with most patient accuracy in experimental research must have had an influence on Davy. Davy was later to be of service to Wedgwood in helping with the publication of his work, his pioneer work, on photography, in the first volume of the *Journals of the Royal Institution* under the title, 'An account of a Method of Copying Paintings upon Glass, and of making Profiles by the Agency of Light upon the Nitrate of Silver'. As one reads today of this process invented by Thomas Wedgwood one is startled, partly by Davy's lack of perception as to the possibilities of the invention, and partly by a realization of the hazards attached to the survival of any thought or fancy, any invention, or any made thing in any medium whatsoever in the race of time. A few may remember Thomas Wedgwood as, in vague terms, the first photographer, although he could not 'fix' his pictures. But it is in connexion with the lives of others, in his discriminating generosity, in his taste and judgment that he really survives – and because his appearance impressed Wordsworth. 'His calm and dignified manner, united with his tall person and beautiful face, produced in me an impression of sublimity beyond what I ever experienced from the appearance of any other human being.' So Wordsworth wrote of Thomas Wedgwood, and Wordsworth was not a man readily enthusiastic over the person or talents of others.

Thomas Wedgwood, as a subscriber to the Pneumatic Institution, saw a good deal of Davy. He tried the airs; but his witness to the powers of nitrous oxide is disappointing, though he did become as it were entranced and, after he had thrown the silk bag from him, kept breathing on furiously with an open mouth and holding his nose with his left hand, having no power to take it away. It is typical of him that, although he was powerless to take his hand away, he was aware of the ridiculousness of his situation. He knew no ecstasy.

The charm of Wedgwood's elegance must have pleased Davy, who liked elegance in others, though he had too much gusty force to be elegant himself. A quality quite other than elegance drew him to another friend made in his Bristol days – Tom Poole of Nether Stowey. It never seems quite fitting to shorten Thomas Wedgwood's Christian name to Tom, though many of his friends did, or to lengthen Tom Poole's name to Thomas. Tom Poole is plain Tom Poole who gloried in the name of tradesman, or as we should now say, business man. He was a partner in his father's tanning yard at Nether Stowey, and later the enterprising owner of it. As well as being a tanner he was a farmer. Of the many good stories of Poole told by Mrs Sandford in her biography[1] – she was a daughter of Gabriel Stone Poole, Tom Poole's cousin, and I can never read her book without wishing we had even half as good a biography of Beddoes – I like two especially. In the first of these Poole is a young man, in the second growing old. In the year 1792 John Poole, a greatly loved uncle of Tom Poole, an able and likable man living at Over Stowey, sickened and died of a malignant fever. His eldest son, also John, who had just gained an Oriel Fellowship, fell ill of the same fever and was very near death. Mrs Sandford writes:

> The only remedy used seems to have been large doses of opium, and these John Poole, a tall powerful young man of two and twenty, and suffering from severe delirium, attended by such spasms in the throat it was feared lock-jaw would ensue, resisted so violently that even his mother was for giving up in despair. These were not the days of trained nurses, and no one knew how to act. But, 'What!' said Tom Poole, 'Let a fine young man *die* for want of a little resolution?' Whereupon, calling in the strongest men from the farm he administered the required medicine by main force.

The second story looks forward to the year 1837 when, at a dinner party, Poole boiled over with rage on hearing a person of local importance disparage in authoritative tones Wordsworth and Coleridge. Poole told the reviler in the most emphatic manner that he was a fool. Mrs Sandford continues:

> A peacemaking friend followed him into the garden, where he found him fuming and panting and pacing up and down, and remonstrating with him for his unwarrantably rude behaviour, soon brought him round to a state of extreme penitence.
> '*Did* I call him a fool? How very wrong of me, how very wrong! Would it be any good to apologise? I am sure if it would give him any satisfaction I would apologise in a moment.'
> So into the house he returned, and at once began to express his regrets.
> 'I am sure, Sir, I am very sorry. I am *very* sorry I was so rude to you just now. I apologise most sincerely. I wish I wasn't so hasty. It

[1] *Thomas Poole and His Friends*, by Mrs Henry Sandford, 2 vols., 1888.

was extremely wrong of me. But – but – but' (with a great gulp, as if he were all but choking) – 'how *could* you be *such a damned fool?*'

As characters, and in their minds, Poole and Davy were contrasts. Poole was plain and tenacious, with something stable in his nature which made him a pillar of strength to those in his neighbourhood less fortunate than himself; though he could, no doubt, as Southey put it, go clod-hopping over their feelings. He was slow-moving where Davy was quick; patient where Davy was irritable. Davy would see what Poole was driving at almost before Poole had begun to say it. It was while Davy was in Bristol that he first visited Poole in his pleasant house, with the well-appointed book-room which we come to know so well. Communication between Bristol and Nether Stowey was remarkably close before the days of quick traffic. Coleridge would walk from Bridgwater, though Poole sometimes provided for him a tolerably meek horse. Davy never made anything of distance all his life; he would go a couple of hundred miles for a bit of fishing.

Poole was twelve years older than Davy and outlived him by eight years. He wrote one of the most balanced eulogies of Davy. Davy, on his part, not long before his death, dedicated his *Consolations in Travel* to Poole with the inscription, 'To Tom Poole, in remembrance of thirty years of continued and faithful friendship.'

Poole was essentially a moderate man; quite other was Robert Southey in his youth, and Southey was more intimate with Davy than Poole when Davy first came to Bristol. Davy must have possessed the Cornish gift, or curse, of adaptability in full measure to have been able to get on with such a varied collection of people as he met at the Beddoes'. Southey called him 'an unreplaceable companion', praised his poetry, and included some of it in *The Annual Anthology* printed in Bristol in 1799. It was Southey who remarked later that Davy would have excelled in any department of art or science to which he had directed the powers of his mind; that he had all the elements of a poet and only wanted the art. (Coleridge said roundly that if he had not been the greatest chemist he would have been the greatest poet of his age.) Southey encouraged Davy to try an epic in the manner of his own long poems; he even proposed a collaboration, but Davy rejected the scheme. Instead Davy planned and partly wrote a poem having Moses as its central figure. John Davy gives a verbatim copy of the plan and some specimens of the composition, in the *Memoir* of his brother which he published as a Prelude to the *Works*. I think *Moses* is the clearest indication we have of what made such a keen judge of men as Coleridge consider Davy a likely poet. The plan precedes de Vigny's poem on the same Biblical figure by many years. Davy was not so much attracted by the mature Moses, Moses the great law-giver, who laments in de Vigny's poem his isolation from ordinary men, as by the young Moses whom he saw as a Jacobin. His notes on the characters throw light on his own ideas at that time. He writes of them, 'Moses a great but enthusiastic Man. Zippora his Superior in Reasoning Powers and in Sensibility. Pharaoh a Despot. Jethro a Wonder, a Philosophic Priest. Joshua a Hero, i.e. a murderer. Miriam the Prophetess, the Sister of Moses, a wonderful Woman.' Davy envisaged Jethro as 'a man of energy' as well as 'that wonder, a Philosophic Priest'. Perhaps he had the aged Unitarian, Dr Priestley, in mind; or perhaps the young Unitarian, S. T. Coleridge. Moses he saw as 'having a light of glory surrounding his body, and under the immediate inspiration of the Deity'. It was Pharaoh, the companion of his youth, to whom, when Moses returned to Egypt after his pastoral life in the desert, he disclosed his 'Jacobinical Sentiments'. The most interesting note is that on Joshua, 'Joshua, a Hero, i.e. a murderer'. In it Davy shows how early in life he formed an idea which was to abide with him. He never, at any time, glorified Napoleon as, say, Hazlitt did. He saw him as Tolstoy was to see him. In a note written at about this time we see, too, how a hope other than that of his revolutionary friends of this period was beginning to beckon him. He wrote:

> Shall those arts which have discovered a thousand instruments for inflicting pain and suffering on civilised man never discover any new means of making him happy? Shall the fruit of the tree of knowledge always continue bitter; shall it never be ripened by the radiance of the sun of benevolence? If there be any sufficiently hard-hearted to believe this, let them remain idle. To us hope, which though it should be vain, is yet an eternal source, will remain; it will ever prompt to actions which though they should deserve no laurels of triumph from mankind, will never have raised them by watering the earth with blood.

In *Consolations in Travel* he wrote that he found desire for universal domination an even more detestable lust than cruelty.

Davy's letters to Southey have vanished, but those letters of Southey to Davy which remain are among Southey's best – and he is an un-commonly good letter-writer. When, in April 1800, he left Bristol for a sojourn in Portugal, he wrote Davy, so much interested in dreams, and half a doctor, an account of a nightmare he had had. The letter anticipates the easy prose of the future biographer of Wesley and Nelson. In Southey's letters to Davy one feels constantly the good prose-writer and the all-too-pedestrian poet. He gives Davy a receipt for an epic as if it were a pie; he thought Davy's proposed poem on Brutus a good subject for a Cornishman. But perhaps it was Joseph Cottle's essay in the epic art which prevented Davy from emulating Southey in the production of such poems as *Thalaba the Destroyer*, which he saw through the press for his friend. Joseph Cottle would certainly be a warning to any young man of intelligence how not to write verse. There are few things funnier in existence than Joseph Cottle's flirtations with the Muse, his *Ascent of the Malvern Hills*, his *Alfred, an Epic Poem*, and his play. Even Southey trembled for Cottle's *Alfred*. In the same letter to Davy in which he recounted his nightmare he wrote:

I tremble for *Alfred* – those long speeches! and, if reviewed by a hostile or even an indifferent hand. Had he listened to me, cut out his dialogues and introduced machinery, he would have done well. Angel and devil nature he would have known as much about as his neighbour, but of human feelings he knows nothing, and might as well write an account of the moon and the history of man on it.

Perhaps the writing of bad verse induces long life, for the writer has all the fun of thinking himself a poet, without the pains. Joseph Cottle outlived most of the young friends of those fervid Bristol days, and wrote his reminiscences of them, so he may be said to have laughed last. He gives us a portrait of Humphry Davy shortly after his arrival in Bristol:

I was much struck with the intellectual character of his face. His eye was piercing, and when not engaged in converse, was remarkably introverted, amounting to absence, as though his mind had been pursuing some severe train of thought scarcely to be interrupted by external objects, and, from the first interview also, his ingenuousness impressed me as much as his mental superiority.

Perhaps, when his eye was so remarkably introverted Humphry, like Lamb, was trying not to laugh. As a pendant to Cottle's description may be placed Southey's description of Davy at about this same period:

> I was in most frequent and familiar intercourse with Davy, then in the flower and freshness of his youth. We were within an easy walking distance of each other, over some of the most beautiful ground in that beautiful part of England. When I went to the Pneumatic Institution he had to tell me of some new experiment or discovery and of the views which it opened for him, and when he came to Westbury there was a fresh bit of *Madoc* for his hearing. Davy encouraged me by his hearty approbation during its progress; and the bag of nitrous oxide with which he usually regaled me upon my visit to him was not required for raising my spirits to settled fair and keeping them at that elevation.

The conclusion of one of Southey's last letters to Davy is:

> Times have changed since we first became intimate, and we also must be changed, you probably more than me, for mine are older and riper habits. I do not love to think of this; the world cannot mend the young man whom I knew before the world knew him, in the very spring and blossom of his genius and goodness.

While he lived in Clifton Davy did not need to breathe nitrous oxide in order to experience a feeling of transport. What saved him from the fatuousness of Cottle was the honesty of his response to the beauty of the visible world, a response which deepened rather than diminished as he grew older. His was not just the worship of nature which was part of the fashion of the time. 'Oh, the evergreen! the evergreen!' One comes much more closely in touch with Davy's exalted moods in his prose than in his stiff verse. In the following passage he relates an experience which Keats later was to feel and express. Davy wrote of his sense of union with all living things:

> Today, for the first time in my life, I have had a distinct sympathy with nature. I was lying on the top of a rock to leeward; the wind was high, and everything in motion; the branches of an oak tree were waving and murmuring in the breeze; yellow clouds, deepening to grey at the base, were rapidly floating over the western

hills; the whole sky was in motion; the yellow stream below was agitated by the breeze; everything was alive, and myself part of the series of visible impressions; I should have felt pain in tearing a leaf from one of the trees . . .

Deeply and intimately connected are all our ideas of motion and life, and this, probably, from very early association. How different is the idea of life in a physiologist and a poet!

It was in the society of Anna Beddoes, sister of Maria Edgeworth, that Davy's enthusiasm was most untrammelled. The strain of admiration and gratitude in which he first wrote of her to his mother was not forced. She combined wit with good humour and good nature. No doubt she was the prototype of the 'wonderful woman' Davy had in mind to draw as the sister of Moses. It was Anna Beddoes who often accompanied Davy on his walks, since Beddoes himself, because of his multitudinous occupations and his bulk, could not walk far. But for at least part of Davy's time in Bristol it is clear, from a letter of Gregory Watt, who also praised Anna's lively mind and happy nature, that she was ill, even dangerously ill. And although her 'simplicity' was authentic, and in contrast to the romantic sham idea of simplicity current at the time, there was, perhaps, in her, a latent tendency to be morbid. 'Feeling' was the fashion. Davy, while at Clifton, was getting up early to write tales with such typical titles as *The Lover of Nature* or *The Feelings of Eldon*; *The Child of Education, or the Narrative of W. Morley*; *The Dreams of a Solitary*; *Imla, the Man of Simplicity, a Romance*. He even tried to emulate Beddoes' success with *Isaac Jenkins, a Moral Tale* by planning *The Villager; a Tale for the common People, to prove that great Cities are the Abodes of Vice*, etc. Vivid and various and gay, Anna Beddoes was also capricious. To John King, Anna's brother-in-law, a Swiss by birth, who practised as a surgeon at Clifton for over fifty years, and who worked as a physiologist with Davy at the Pneumatic Institute, Davy remarked in a letter soon after he left Bristol for London:

How I should rejoice to hear that Mrs Beddoes might be called by the sacred name of mother! How delightful a thing it would be to see that woman of genius, of feeling, of candour, of idleness, of caprice, nursing and instructing an infant, and losing in one deep sympathy many trifling hopes and fears.

In addition to praising her charm and liveliness he felt the poetry of her nature. She possessed, he said, 'a fancy almost poetical in the highest sense of the word, great warmth of affection, and disinterestedness of feeling, and, under favourable circumstances she would have been even in talents, a rival of Maria'.

Anna Beddoes was to have four children, two boys and two girls. The elder boy was Thomas Lovell Beddoes, who grew from being a child a little like Hartley Coleridge – a child innocently gay, with a gibe always on his tongue, a mischievous eye, and locks curling like the hyacinth – into a strange and wandering man, the author of *Death's Jest Book*, a poet who is much more than a shadow of the Elizabethans, a poet who vies with the greatest when occasion holds him to it. What if Davy in the pride of life could have read the words to be written by this poet yet unborn:

> the world is open:
> I wish you life and merriment enough
> From wealth and wine, and all the dingy glory
> Fame doth reward those with, whose love-spurned hearts
> Hunger for goblin immortality.
> Live long, grow old, and honour crown thy hairs
> When they are pale and frosty as thy heart.

That Anna Beddoes was not merely a woman whom Davy chose to be romantic about, but one whom he remembered always with gratitude for her grace and kindness, is clear from the verses he addressed to her from Ireland eight years after he had left Bristol. From verses signed A. B. and copied by Davy into a notebook, it appears that Anna herself was a little in love with him. But it was her nature to love and help all those who came to the house. The letter which reveals her most completely in her gentleness and diffidence, and which also shows Dr Beddoes as other than the bustling theorist one sometimes thinks him, was addressed to Thomas Wedgwood. It offered him hospitality and help in his sickness and concluded:

The Dr, you know, is a most peaceable being, and could not disturb you. I would not, and I know nothing that would gratify me more than to nurse you whenever you would suffer me to do anything for you; indeed, in this respect, you would find me another sister, though I am aware of my vanity in saying so. I will not fatigue you by adding anything more than my name, A. M. Beddoes.

5 · Davy and Coleridge

The most fruitful friendship Davy made with a poet was with S. T. Coleridge. *Lyrical Ballads* had been published in the summer of 1798, and Wordsworth and Coleridge had left Somerset for their travels in Germany in the early autumn, just before Davy came to Clifton. It was not until after Coleridge's return from Germany that his friendship with Davy began – when he revisited Bristol and Nether Stowey, and when Davy visited London.

The two had heard a good deal of each other from Southey. In December 1799, when Coleridge wrote to Southey suggesting that Southey should join him in London, he qualified his invitation with, 'This I should press on you were not Davy at Bristol – but he is indeed an admirable young man, not only must he be of comfort to you, but on whom can you place such reliance as a medical man?' Towards the end of the same letter he wrote, 'I am afraid that I have scarce poetic Enthusiasm enough to finish *Christabel* – but the poem with which Davy is so much delighted, I probably may finish time enough.'

That same December Davy paid a visit to London, and he and Coleridge dined with Godwin. A day or two before, small Hartley Coleridge had given Mr Gobwin, as he called the author of *Political Justice*, such a rap on the shins with a ninepin that 'Gobwin', in huge pain, had lectured Sara Coleridge on her son's boisterousness. Coleridge had agreed that his child was somewhat too rough and noisy; he agreed until he and Davy dined with the Godwins. Then he found the silence of Godwin's children quite catacombish and, thinking of Mary Wollstonecraft (she had died on September 10th, 1797, after marrying Godwin on March 29th of the same year), was oppressed by it.

Davy admired Godwin, and must have been flattered to receive the attention of one who was so much the cry among clever young men. Godwin praised Davy, and this praise Coleridge joyfully repeated both to Davy himself and to Thomas Wedgwood. 'I like him [Godwin]', he wrote to Wedgwood, 'for thinking so well of Davy. He talks of him everywhere as the most extraordinary human-being he has ever met with.' Coleridge adds, 'I cannot say that, for I know one whom I think to be the superior – ; but I never met so extraordinary a *young* man.'

In the New Year the extraordinary young man received one of Coleridge's flying letters, a letter which has survived all vicissitudes, although there are a number of holes burned, as though by drops of acid, in the original now in the possession of the Royal Institution. Coleridge had not yet entirely abandoned his idea of founding a little colony of kindred spirits on the banks of the Susquehanna or elsewhere; Davy is being wooed, half-playfully, as a likely associate:

. . . Davy! Davy! If the public good did not iron and adamant you to England and Bristol, what a little colony might we not make. Tobin, I am sure, would go, and Wordsworth, and I, and Southey. Precious stuff for Dreams – and God knows I have no time for them! . . .

A Private Query – On our system of Death does it not follow that killing a bad man might do him a great deal of good? And that Buonaparte wants a gentle Dose of this kind, dagger or bullet *ad libitum*? I wish in your Researches that you and Beddoes would give a compact compressed History of the Human Mind for the last century, considered simply as the acquisition of Ideas or new arrangement of them. Or if you won't do it there, do it for me – and I will DO it with an Essay I am now writing on the principles of Population and Progressiveness.

Godwin talks evermore of you with lively affection. – 'What a pity that such a man should degrade his vast Talents to Chemistry,' cried he to me – 'Why,' quoth I, 'how Godwin! can you thus talk of science, of which neither you nor I understand an iota' etc., and I defended Chemistry as knowingly at least as Godwin attacked it – affirmed that it united the opposite advantages of immaterialising the mind without destroying the definiteness of the Ideas – nay even while it gave clearness to them – and eke was being necessarily performed with the passion of Hope, it was poetical and we both agreed (for Godwin as well as I thinks himself a Poet) that *the Poet* is the greatest possible character etc., Modest Creatures! Hurrah, my dear Southey! – and you and I and Godwin and Shakespeare, and

Milton, with what an athanasiophagous Grin we shall march together – *we poets*: Down with all the rest of the World! By the
word athanasiophagous I mean devouring Immortality by anticipation! 'Tis a sweet word!

God bless you, my dear Davy! Take my nonsense like a pinch of
snuff – sneeze it off, it clears the head – and to Sense and yourself
again – With the most affectionate esteem

Your's ever,

S. T. COLERIDGE

By May 1800, when Coleridge was staying with Poole at Nether
Stowey, he wrote to Godwin:

> In Bristol I was much with Davy – almost all day. He always
> talks of you with great affection, and defends you with a friend's
> zeal against the Animalcula, who live on the dung of the great
> Dung-fly Mackintosh. If I settle at Keswick, he will be with me in
> the fall of the year and so must you – and let me tell you, Godwin!
> four such men as you, I, Davy, and Wordsworth, do not meet
> together in one house every day in the year – I mean, four men so
> distinct with so many sympathies.

While still at Nether Stowey Coleridge wrote Davy the following
letter – a paragraph giving particulars of the contents of Blumenbach's
Manual is here omitted:

> Saturday Morning, Mr T. Poole's, Nether
> Stowey, Somerset.

> My dear Davy . . . I received a very kind letter from Godwin, in
> which he says he never thinks of you but with a brother's feeling of
> love and expectation. Indeed, I am sure he does not.

> I think of translating Blumenbach's Manual of Natural History:
> It is very well written, and would, I think, be useful both to students
> as an admirable direction to their studies, and to others it would
> supply a *general* knowledge of the subject . . . I have the last
> edition, i.e., that of April, 1799. Now, I wish to know from you
> whether there is in English already any work of one volume (this
> would make 800 pages), that renders this useless. In short, should
> I be right in advising Longman's to undertake it? Answer me as soon
> as you conveniently can. Blumenbach has been no great discoverer,

[59]

though he has done some respectable things in that way, but he has enormous knowledge and an *arranging* head. Ask Beddoes, if you do not know.

When you have leisure you would do me a great service, if you would briefly state your metaphysical system of impressions, ideas, pleasures, and pains, the laws that govern them, and the reasons which induce you to consider them as essentially distinct from each other. My motive for this request is the following: — As soon as I settle, I shall read Spinoza and Leibnitz, and I particularly wish to know wherein they agree with, and wherein they differ from you. If you will do this, I promise you to send you the result, and with it my own creed.

God bless you!

S. T. COLERIDGE.

In a postscript Coleridge added:

Blumenbach's book contains references to all the best writers on each subject. My friend T. Poole, begs me to ask what, in your opinion, are the parts or properties in the oak which tan skins? and is cold water a complete menstruum for these parts or properties? I understand from Poole that nothing is so little understood as the chemical theory of tan, though nothing is of more importance in the circle of manufactures; in other words, does oak bark give out to cold water all those of its parts which tan?

To this letter, and to another note from Coleridge which he had received, Davy answered immediately. The five pounds referred to had been borrowed by Coleridge from Davy some time before; the 'acid' was a remedy of Davy's in which Coleridge had such faith that he wished Davy to send some to his mother-in-law.

My dear Coleridge[1]

I received your letter and the five pounds on Thursday evening and I sent a notice of it with the parcel of acid to Mrs Fricker. It was not sent till early this morning; but perhaps you will receive it before this letter. I am sorry your resolution is fixed on the northern dwelling. I am *even* disappointed, for the omniscient Mr B. Coates

[1] This letter has come only recently into the possession of the Royal Institution where I was kindly permitted to copy it.

told me that he had heard from *good authority* that you had taken or were about to take a house at Stowey. I have talked with Beddoes about Blumenbach. He says that there is no such work in English. He considers it as a good work and a useful work; but agrees with me in thinking that the loss of time and waste of energy in translation of it would be badly bestowed by a poet philosopher. You were born to connect man with Nature by the intermediate links of harmonious sounds and to teach them to disconnect this feeling from unmeaning words. I am certain that with regard to profit you will get more by writing originally than by writing these confounded translations of German books. I will write as soon as I have leisure on the outline of my metaphysical notions, if it is possible to give sufficient consistence to theories which have been constantly altering and undergoing new modification. But before I have leisure, I hope to see you here. Recollect your promise to return in a week.

The french [*sic*] have lately fully investigated the subject of tanning. Séguin has discovered that the oak bark and other barks owe their properties to two principles, both of which are soluble to cold and hot water. The tanning principle which he has been able to exhibit in a separate state and has called tannin and the gallic acid or astringent principle.

I believe he says that the conversion of skin into leather depends upon its combination with these two principles. The paper is one of the last No. of the Annales de Chimie and I *believe* in Nicholson's Journal. If I can get the number I will send it to Mr Poole.

farewell yours
with warm affection

H. DAVY

Please remember me to Mrs Coleridge.

Monday afternoon – I have removed my furniture into the garden amidst the strawberries and am now writing under the shade of an apple tree thus I begin to claim a relationship with nature.

The next letter which has survived from Coleridge to Davy was written in July. Davy's hope that Coleridge might settle at Nether Stowey was not realized; Coleridge had, in Lamb's words, gone to

join his god, Wordsworth, in the North. Davy had lent him a little
money to help get him there. Coleridge wrote from Grasmere:

Wed. July 15, 1800.

My dear Davy,[1]

   . . . I hope, that you have suffered no inconvenience from want
of the money, which I borrowed of you – it has made me very
uneasy; but in a few days I will take care, that it shall be remitted
to you. We remove to our own house at Keswick on Tuesday week –
my address is, Mr Coleridge, Greta Hall, Keswick, Cumberland.
My dear fellow, I would that I could wrap up the view from my
House in a pill of opium, and send it to you! I should then be sure of
seeing you in the fall of the year. But you *will* come. –

   As soon as I have disembrangled my affairs by a couple of months'
Industry, I shall attack chemistry, like a Shark –. In the meantime
do not forget to fulfil your promise of sending me a synopsis of your
metaphysical opinions. I am even *anxious* about this. – I see your
Researches on the nitrous oxyde regularly advertised – Be so kind as
to order one to be left for me at Longman's, that it may be sent in
my box.

After a detailed description of the symptoms of his illness Coleridge
concluded:

   I have read the little chemist's pocket book twice over. – Do, do,
my dear Davy! come here in the fall of the year. – Sheridan has
sent to me again about my Tragedy – I do not know what will
become of it – he is an unprincipled Rogue.

   Remember me to Mr Coates when you see him – and be sure you
do to Mathew Coates and to Mrs Coates. Will you be so kind as just
to look over the sheets of the lyrical Ballads? What are you now
doing? – God love you! Believe me most affectionately, my dear
Davy, your friend

S. T. COLERIDGE

The reference to the *Lyrical Ballads* is to a second edition which
Wordsworth and Coleridge were contemplating. The intention was to
include many poems of the original edition, together with a second
volume of new poems in which it was first intended that *Christabel*, so
much admired by Davy, should have a place. Wordsworth wrote to

---

[1] S. T. C., *Letters*, i. 604.

Davy and Coleridge

Davy, whom he had not yet met, but who was so conveniently near the premises of the printers, Biggs and Cottle of St Augustine's Back:

> You would greatly oblige me by looking over the enclosed poems and correcting anything you find amiss in the punctuation a business at which I am ashamed to say I am no adept . . . I write to request that you would have the goodness to look over the proof-sheets of the 2nd volume before they are finally struck off. In future I mean to send the Mss to Biggs and Cottle with a request that along with the proof-sheets they may be sent to you . . . Be so good as to put the enclosed Poems into Mr Biggs' hands as soon as you have looked them over in order that the printing may be commenced.

So it came about that Davy, still in the freshness of his youth, opened the wrappers of successive packets and read, in their first freshness, sometimes in Dorothy Wordsworth's handwriting, sometimes in Coleridge's, sometimes in Coleridge's corrected by Wordsworth, the poems which go to make up the 1801 edition of the *Lyrical Ballads*. He must have been one of the first to read, 'A slumber did my spirit seal', 'She dwelt among the untrodden ways', and *Michael*.

He read also *A Poet's Epitaph*. In the account of Davy which J. Cordy Jeaffreson[1] cooked up out of J. A. Paris many years after Davy's death he implies that Wordsworth's disappointment with Davy was bitterly expressed in his stanza warning the 'philosopher' to take his dwindling soul away from the poet's grave. But when Wordsworth wrote that line Davy was not twenty-one and he and Wordsworth had never met. The conjuration of the doctor and the philosopher was in general. If a particular doctor – and Beddoes, if not a rosy man, was 'right plump to see' – and his young experimentalist ever dreamt of applying Wordsworth's lines to themselves, it would only have added a little spice to their fun as they both parodied one or two of the more tempting lyrical ballads. Davy's effort is mingled with accounts of galvanic experiments in a notebook labelled, 'Clifton 1800 from August to Novr.' Extricated from various other jottings it reads:

> As I was walking up the street
> In pleasant Burny town
> In the high road I chanced to meet
> My Cousin Matthew Brown.

[1] *A Book About Doctors*, J. Cordy Jeaffreson, 1860, vol. i, pp. 68–72.

The Mercurial Chemist

My Cousin was a simple man
A simple man was He
His face was of the hue of tan
And sparkling was his eye —

His coat was red for in his youth
A soldier he had been,
But He was wounded and with ruth
He left the camp I ween —

His wound was cured by Doctor John
Who lives upon the hill
Close by the rock of grey free stone
And just above the mill.

He then became a farmer true
And took to him for aid
A wench who though her eye was blue
Was yet a virgin maid.

He married her and had a son
Who died in early times
As in the churchyard is made known
By poet Wordsworths Rymes.

As long as this fair wife did prove
To him a wife most true
His red coat He away did shove
And wore a coat sky blue.

This was fun, of course. Davy shared some faculty of soul with Wordsworth; but he was not prostrate with adoration. Yet, although Davy was quick and comprehensive where Wordsworth was narrow and profound, they were more like each other than either was like Coleridge. They shared a common-sensical basis. Both jumped from a kind of ordinariness, the one to the glorious summits of poetry, the other to an intuitive perception of how things worked. But in Coleridge there was nothing ordinary. He was all in all astonishing. Davy tried to hearten him, to combat his self-distrust, to keep alive his faith in his power to finish the second part of *Christabel*. But after much

Market Jew Street, Penzance, c. 1831

Davy at Clifton, *c.* 1800. A pastel portrait by James Sharples of Bath

wavering on the part of Wordsworth and Coleridge the poem was not included in the edition of the *Lyrical Ballads*. Could Davy's influence over Coleridge have been stronger at this time *Christabel* would never have been allowed to remain much read privately, imitated, but unpublished until 1816.

The contrast in character between Davy and Coleridge is striking – Coleridge with his subtle mind, his shaping spirit of imagination, and his practical weakness; and Davy in whom imagination was made to serve utility, but in whom was candour, sympathy, a Baconian splendour of vision, and an organic sensibility as excitable as Coleridge's own. In Cornwall he had been moved to write verse by the sight of the moonlight streaming on St Michael's Mount; Clifton found him going to see Tintern Abbey by moonlight; walking along the Avon by moonlight; finding always in the full moon a wildness in his mind, a kind of indefinite sensation more usually associated with the poetical than the scientific temperament.

Coleridge loved Davy. He is glad 'with a stagger of the heart' to see his handwriting. He writes to him of his great schemes, how he is going to write a book concerning the affinities of the feelings with words and ideas under the title 'Concerning Poetry and the Nature of the Pleasures derived from it'. He dilates on his scheme and then says: 'To whom shall a young man utter his *pride* if not to the young man whom he loves.' He wants to climb the hills with Davy. Writing from Keswick, he says he can see mountains from his window, a great camp of mountains. 'Each mountain is a giant's tent, and how the light streams from them!' He says he aches for Davy to be with him and Wordsworth, and says Wordsworth is a lazy fellow for not having written Davy about the proof-sheets. Coleridge calling Wordsworth a lazy fellow! Coleridge wants to be a chemist like Davy so as to sympathize with him from the middle of his heart's heart. He is going to fit up a little lab. with William Calvert, and Davy must advise them about apparatus. Calvert already has 'an electrical machine and a number of little nicknacks connected with it'. This is in a postscript.

Davy must have smiled over the nicknacks, and over the descriptions of Coleridge's impetuous leap from Parnassus into the jungles of science. Yet the influence of his friend's remarks on the limitations of chemistry is apparent in Davy's work. Coleridge later defined philosophy as the science of ideas, and science as the knowledge of powers. But he was never a scientist in embryo as, perhaps, Shelley was.

The Mercurial Chemist

Coleridge wrote: 'As far as words go I have become a formidable chemist – having got by heart a prodigious quantity of terms, etc., to which I attach some ideas, very scanty in number I assure you, and right meagre in their individual persons. That which must discourage me in it is, that I find all power of vital attributes to depend on modes of arrangement, and that chemistry throws not even a distant rushlight glimmer upon this subject.' He becomes more and more interested in chemistry, however; but then he reflects that his passion for science is scarcely true or genuine, 'it is but Davyism; that is, I fear that I am more delighted at your having discovered facts than at the facts having been discovered'. Characteristically he hoped more proudly of Davy than of himself. He once said that he hoped more proudly of Davy than of any other man.

As there are more letters preserved from Coleridge to Davy than from Davy to Coleridge we see Coleridge the more clearly of the two. But a letter always throws light on the person to whom it is written as well as on the person writing it, and Davy must have been capable of loving Coleridge in the light of knowledge to have received such letters. He perceived and analysed Coleridge's weakness; he was not I think quite so warmhearted towards Coleridge as Coleridge towards him. But he greatly admired Coleridge and knew that his own spirit had been quickened by him. As to Coleridge's opinion of Davy, we have it summed up in his own expressive words to Cottle who had asked him what he thought of Davy in comparison with the other talented men of London. Coleridge replied: 'Why, Davy can eat them all! There is an energy, an elasticity in his mind, which enables him to seize on and analyse all questions, pushing them to their legitimate consequences. Every subject in Davy's mind has the principle of vitality. Living thoughts spring up like turf under his feet.'

Between 1799 and 1802 Coleridge was writing his leaders for the *Morning Post*. His attitude towards the French war and towards Pitt had changed since the days when he had first lectured in Bristol, talked joyous treason, produced *The Watchman* and written *Fire, Famine and Slaughter*. Now he fought Napoleon. He was intensely alive to political ideas. So was Davy. But there was an event in 1800 which influenced Davy's life far more than any political happening. In that year Volta invented a crude electric battery, the 'voltaic pile', and in that same year Carlisle and Nicholson resolved water into oxygen and hydrogen by means of an electric current. On July 3rd, 1800,

Davy and Coleridge

Davy wrote in a letter to Davies Gilbert – 'We [meaning himself and Beddoes] have been repeating the galvanic experiments with success.' Immediately he instituted further experiments of his own and began the long train of researches which led to his most fortunate discovery in 1807. In his different way Coleridge was as much interested in galvanism as Davy. He loved, when in Bristol, to watch Davy at work; he reveals in his letters how much the world of the experimentalists impressed itself on his imagination, and how eager he was to read anything Davy might publish in *Nicholson's Journal*. In October 1800 he wrote to Davy:

> Many a moment have I had all my France-and-England curiosity suspended and lost, looking in the advertisement front-columns of the Morning Post Gazeteer, on *Mr. Davy's Galvanic Habitudes of Charcoal*. Upon my soul, I believe there is not a Letter in those words, round which a world of imagery does not circumvolve; your room, the Garden, the cold bath, the Moonlit Rocks, Barrister Moore, and simple-looking Frere and dreams of wonderful Things attached to your name – and Skiddaw, and Glaramara, and Eagle Crag, and you, and Wordsworth, and me on the top of them!

Coleridge filled up a blank space in this letter with his amusing *Skeltoniad* on Mackintosh, in which he apostrophized Davy, 'Ho! Ho! Brother Bard.' He also said that Wordsworth was fearful Davy had been much 'teized' by the printers on his account. He had been indeed.

For the most part Coleridge's letters to Davy at this time are in festival humour. But the letter of December 2, 1800, shows that Davy himself was not well and that, significantly, Coleridge is interested in the subject of pain:

> Greta Hall, Tuesday Night, Decemb. 2, 1800.
>
> My dear Davy,
>
> By an accident I did not receive your Letter until this Evening. I would, that you had added to the account of your indisposition the probable causes of it. It has left me anxious, whether or no you have not exposed yourself to unwholesome influences in your chemical pursuits. There are *few* Beings both of Hope and Performance but few who combine the 'Are' and the 'will be' – For God's sake, therefore, my dear fellow, do not rip open the Bird, that lays the golden Eggs. I have not received your Book – I read yesterday a sort of Medical Review of it. I suppose, Longman will send it to me

when he sends down the Lyrical Ballads to Wordsworth. I am solicitous to read the latter part – did there appear to you any remote analogy between the case, I translated from the German Magazine, and the effects produced by your gas? – Did Carlisle[1] ever communicate to you, or has he in any way published, his facts concerning *Pain*, which he mentioned when we were with him? It is a subject which *exceedingly interests* me – I want to read something by somebody expressly, on *Pain*, if only to give an *arrangement* to my own thoughts, though if it were well treated, I have no doubt it would revolutionize them . . .

In December Coleridge writing to Poole said, 'You have conversed much with Davy – he is delighted with you. What do you think of him? Is he not a great Man, think you?' Coleridge's last letter to Davy himself, while Davy was still at Clifton, answered a remark of Davy's about Wordsworth's *Michael*. Davy had said that the poem 'was full of just pictures of what human life ought to be'. Coleridge replied on January 11th, 1801:

> You say W's 'last poem is full of just pictures of what human life ought to be' – believe me, that such scenes and such characters really exist in this county – the superiority of the small Estatesman such as W. paints in old Michael, is a God compared to our peasants and small Farmers in the South: and furnishes important documents of the kindly ministrations of local attachment and hereditary descent –
>
> Success, my dear Davy! to Galvanism and every other ism and schism that you are about. Perge dilectissime! et quantum potes (potes autem plurimum) rempublicam humani generis juva. Videtur mihi saltem alios velle – te vero posse. Interea a Deo optimo maximo iterum atque iterum precor, ut Davy meus, Davy, meum cor, meum caput, mea spes altera, vivat, ut vivat diu et feliciter! –
>
> Tui amantissimus
S. T. COLERIDGE[2]
>
> Raptum properante Γραμματαφόρῳ

[1] Sir Anthony Carlisle (1768–1840), surgeon.

[2] The extracts used here are from the text of this letter as reconstructed in *Collected Letters of S. T. C.*, ed. Earl Leslie Griggs, vol. ii, pp. 373–4. The manuscript is in the Royal Institution. I have omitted the square brackets by which the editor indicates his expansions of S. T. C's abbreviations. Coleridge knew that Dr Cardew's pupil would get his meaning; as also of the last three words scribbled in schoolboy vein, 'dashed off while the postman is approaching'.

6 · *London*

It was natural that Davy, who was hearing so much about education from Dr Beddoes, and of small Hartley Coleridge from Hartley's father, should become, by letter, his mother's collaborator in the bringing up of the younger children at Penzance, counselling her as to the education of his three sisters and his young brother John with the air of a sage old man. Davy never seems more youthful than when he is a little pompous. The family fortunes had improved. A legacy had made Mrs Davy, although she still let part of her house, independent once more. Davy himself was self-supporting, and he relinquished all claims on his father's estate in favour of his mother and the younger children. Debts had been paid. Letters to John Tonkin show that close relations were maintained, and that Davy, in more immediate connexion with the world than his old friends, executed many little commissions for them. But Tonkin must have shaken his head many times over the idea of his hoped-for successor – he had wished that Humphry should succeed him at Penzance – in a nest of Jacobinical rascals. With his sententious maxims, he was utterly opposed to any activities which could be interpreted as subversive. No doubt he talked darkly to Miss Allen about the boy's not being able to touch pitch without being defiled.

The first time Humphry went back to Penzance was in the fall of the year, nearly twelve months after his first going away. His hazardous trials of his body in the experimental inhaling of gases had made a holiday necessary, and he arrived at Penzance before his letter saying he was coming, breaking in on his Aunt Millet at Marazion where the coach-road ended, and then going on to his mother's house in Market Jew Street. He always enjoyed his mother's cooking, especially her marinated pilchards. For a month he ate well, drank wine instead of inhaling nitrous oxide, and breathed that air of West Penwith whose properties he could not experimentally account for. He took a little portable apparatus with him so as to carry on what work he could. Perhaps it was because of John Tonkin's disapproval that he was always so anxious that his family should think he was getting on well.

His advice in his letters as to Kitty, Grace, Betsy, and John shows a good deal of common-sense and a considerable fund of schoolmasterishness. If Kitty will write him a good letter, well written and well spelt, he will write her a long one in return. He thinks locks of hair silly things, but he sends her one. He tells his mother that Kitty should read and write more than work. Still writing of Kitty, he advised his mother to give her history, give her tales, give her anything but commonplace novels, so long as it made her fond of reading. With Anna Beddoes no doubt in mind, he said that elegant education was of little importance; the cleverest women he knew could neither dance, sing, nor play on the harpsichord. He is a little caustic about Betsy's schoolmistress. 'After Betsy has learnt French and English perfectly, I should not trouble myself by keeping her with Miss E. C. Imitation has a great influence, and I would not wish a sister of mine to be like her in everything.' John was evidently being a little wayward and his brother wrote, 'I hope John's hand and head will improve together. Endeavour to make him fond of reading and drawing, or of any amusement that will give him habits of attention. He must be praised for his imperfect performances, and he will improve. The germ of future abilities must be produced by giving him a desire to distinguish himself for something, no matter what.'

Davy was a man of sensibility in the meaning of the word in his own period which the Oxford Dictionary defines as 'capacity for refined emotion; delicate sensitiveness of taste; also, readiness to feel compassion for suffering, and to be moved by the pathetic in literature or art'. Trembling sensibility became an affectation, and so was laughed out of fashion; but when Davy was young he thought innate sensibility a better indication of what a boy might become than quickness of mind or a retentive memory. In one of his Clifton notebooks there is a draft of a letter which he wrote for his young brother John to give to the refugee Frenchman in Penzance who had been his own French teacher. The letter is not dated, and there is no evidence that it was ever sent; but Davy thought it of sufficient importance to make a rough copy before writing a fair one. The letter reads:

My dear Sir,
    The little boy who brings you this letter is my brother. It is my desire and it is his mother's desire that he should become your pupil.

I fear his mind at present is in a very uncultivated state; but he seems to possess sensibility which I have been accustomed to consider as the foundation of all power and activity.

Under your tuition at all events, he must be improved and if he is not capable of becoming learned, he will at least become virtuous.

His disposition at present is good; but his habits are irregular. These habits you will easily correct without pain; at a future period he will thank you – and if indeed the benefits he may derive from you are at all analogous to those which his brother has derived he will never forget his benefactor –

At no time in his life was Davy an amusing letter-writer, though his correspondence was extraordinarily wide. The world of intellectual men was small, and they reached out over long distances to one another, over friendly and even over enemy countries. In general, Davy's letters, even to his mother, are stiff. No doubt he knew they would be passed round – a restraining thought. The letter to his mother, written in November 1800, just two years after he had left home, gives the only direct vignette we have of mother and son in the little parlour in Market Jew Street, projecting their hopes into the future:

Hot Wells, November 10, 1800.

My dear Mother,

Had I believed that my silence of six weeks would have given you a moment's uneasiness, I should indeed have written long ago. But I have been much engaged in my favourite pursuit of experimenting, and in endeavouring to amuse two of my friends who have been staying at the Institute. One of them is your quondam lodger, Gregory Watt, who desired to be kindly remembered to you and the family. The other you have heard me speak of: his name is Thomson; and he is one of the few to whom God has given a spirit carrying them above the common things of the world.

Accept my affectionate thanks for your presents. I have received them all, and I have made good use of them all. Several times has a supper on the excellent marinaded pilchards made me recollect former times, when I sat opposite to you, my dear mother, in the little parlour, round the little table, eating of the same delicious food, and talking of future unknown things. Little did I then think of my present situation, or the mode in which I am, and am to

[71]

be, connected with the world. Little did I then think that I should ever be so long absent from the place of my birth as to feel longings so powerful as I now feel for visiting it again.

I shall see with heartfelt pleasure the time approaching when I shall again behold my first home – when I shall endeavour to repay some of the debts of gratitude I owe to you, to the Doctor, and to my aunts. My next visit shall not be so short a one as my last. I shall stay with you at least two or three months. You have let half your house. Have you a bedroom reserved for me, and a little room for a laboratory? Which part have you let?

When I come to Penzance we will settle all about John; till then I should like him to learn French and Latin with Mr Dugart. The expense of this or any other part of his education I shall be glad to defray. Do not by any means put him with Dr Coryton. I have long procured the paints: if there is no vessel in the course of a week, they shall be sent off by waggon.

I will write to Kitty in the course of the next month. I am glad Grace is better. Remember me with affection to her. I have not yet seen Mr Griffen. Any one who has lately seen my friends I shall be glad to see.

Have the goodness to tell Dr Borlase that I will endeavour to procure the book he wished for in London.

All in the way of progress goes on nobly. My health was never better than it has been since I left Cornwall last. I shall be glad to hear from you soon. You have a hundred objects to write about interesting to me. I can write only of myself. Remember me affectionately to all my friends (particularly the Doctor), my aunts and uncles. Love to Kitty, Grace, Betsy and John.

Farewell, my dear mother,

I am your affectionate son,

H. DAVY.

A letter to John Tonkin, who was nearly eighty-two, was written in the new year succeeding the above:

Dowery Square, Clifton, January 12, 1801.

Respected Sir,

I have sent in the box enclosing this letter and set of paints for John, two bottles, containing different preparations of phosphorous, with directions for using them. The mode of conveyance by the waggon is very slow; I shall not, therefore, attempt to fill my pages with anything that may be called news. Never was the state of public affairs in England more confused than at this moment, and never were the hopes of peace and plenty feebler in the public mind. . . .

I am at the moment very healthy and very happy; I have had great success with my experiments, and I gain a competence by my pursuits, at the same time that I am (in hopes at least), doing something towards promoting the public good. If I feel any anxiety, it is that of being removed so far from you, my mother, and my relations and friends. If I was nearer, I would endeavour to be useful to you; I would endeavour to repay some of the debts of gratitude I owe to you, my first protector and earliest friend. As it is I must look forward to a futurity that will enable me to do this: but, believe me, wherever I am, and whatever may be my situation, I shall never lose remembrance of obligations conferred on me, or the sense of gratitude that ought to accompany them.

> I remain, respected Sir,
> With unfeigned duty and affection, yours,
> H. DAVY.

When, in this letter, Davy wrote 'wherever I am, and whatever may be my situation', he knew of his almost certain removal to London. Of this determination, and of his prospects, he wrote in detail to Davies Gilbert:

> Hotwells, March 8, 1801.

I cannot think of quitting the Pneumatic Institution, without giving you intimation of it in a letter; indeed, I believe I should have done this some time ago, had not the hurry of business, and the fever of emotion produced by the prospect of novel changes in futurity, destroyed to a certain extent my powers of consistent action.

You, my dear Sir, have behaved to me with great kindness, and the little ability I possess you have very much contributed to develop; I should therefore accuse myself of ingratitude, were I to neglect to ask your approbation of the measures I have adopted with regard to the change of my situation, and the enlargement of my views in life.

In consequence of an invitation from Count Rumford, given to me with some proposals relative to the Royal Institution, I visited

London in the middle of February, where, after several conferences with that gentleman, I was invited by the Managers of the Royal Institution to become the Director of their laboratory, and their Assistant Professor of Chemistry; at the same time I was assured that, within the space of two or three seasons, I should be made sole Professor of Chemistry, still continuing Director of the laboratory.

The immediate emolument offered was sufficient for my wants; and the sole and uncontrolled use of the apparatus of the Institution, for private experiments, was to be granted me.

The behaviour of Count Rumford, Sir Joseph Banks, Mr Cavendish, and the other principal managers, was liberal and polite; and they promised me any apparatus that I might need for new experiments.

The time required to be devoted to the services of the Institution was but short, being limited chiefly to the winter and spring. The emoluments to be attached to the office of sole Professor of Chemistry are great; and, above all, the situation is permanent, and held very honourable.

These motives, joined to the approbation of Dr Beddoes, who with great liberality has absolved me from my engagements at the Pneumatic Institution, and the strong wishes of most of my friends in London and Bristol, determined my conduct.

Thus I am quickly to be transferred to London, whilst my sphere of action is considerably enlarged, and as much power as I could reasonably expect, or even wish for at my time of life, secured to me without the obligation of labouring at a profession.

The Royal Institution will, I hope, be of some utility to society. It has undoubtedly the capability of becoming a great instrument of moral and intellectual improvement. Its funds are very great. It has attached to it the feelings of a great number of people of fashion and property, and consequently may be the means of employing, to useful purposes, money which would otherwise be squandered in luxury, and in the production of unnecessary labour.

Count Rumford professes that it will be kept distinct from party politics; I sincerely wish that such may be the case, though I fear it. As for myself, I shall become attached to it full of hope, with the resolution of employing all my feeble powers towards promoting its true interests.

So much of my paper has been given to pure egotism, that I have but little room left to say any thing concerning the state of science, and the public mind in town; unfortunately, there is little to say. I have heard of no important discoveries. In politics, nothing seems capable of exciting permanent interest. The stroke of poverty, though severely felt, has been a torpedo, benumbing all energy, and not irritating and awakening it, as might have been expected.

Here, at the Pneumatic Institution, the nitrous oxide has evidently been of use. Dr Beddoes is proceeding in the execution of his great popular physiological work, which, if it equals the plan he holds out, ought to supersede every work of the kind.

I have been pursuing Galvanism with labour, and some success. I have been able to produce galvanic power from simple plates, by effecting on them different oxidating and de-oxidating processes; but on this point I cannot enlarge in the small remaining space of paper.

Your remark concerning *negative* Galvanism, and de-oxidation, is curious, and will most probably hold good.

It will give me much pleasure to see your mathematical Paper in the Philosophical Transactions, but it will be, unfortunately, to me the pleasure of *blind* sympathy, though derived from the consciousness that you ought to be acting upon, and instructing the world at large.

It will give me sincere pleasure to hear from you, when you are at leisure. After the 11th I shall be in town – my direction, Royal Institution, Albemarle Street. I am, my dear friend, with respect and affection,

Yours,

HUMPHRY DAVY.

In a letter to his mother about his proposed move to London Davy considerably exaggerated his financial prospects. He also wrote, 'You will all, I dare say, be glad to see me getting among the *Royalists*, but I will accept of no appointment except upon the sacred terms of *independence*.' By this he obviously meant political independence.

Count von Rumford, founder of the Royal Institution of Great Britain, and its secretary until he left England in May 1802, was the scientist by whose side Davy had audaciously ranged himself in his first Essays. Rumford's famous cannon-boring experiment, by which he had made the first determination of the mechanical equivalent of heat, had been carried out at Munich.

He was as unlike Dr Beddoes as two men could be. The doctor was short and round; the Count tall and elegant. He was six feet in height and had been handsome. Externally he was a man of fashion, whereas there was always something a little ridiculous in Beddoes. Yet there was an element of the tycoon in Rumford which the little doctor was entirely free of. When Davy joined the Royal Institution the Count, although not quite fifty, had been through a series of adventures which might make him the hero of a picaresque novel, with the Rumford roaster as his device. His dominating desire was to make science useful. Where the doctor wished to remove pain of body, the Count wished to remove nuisances from houses, nuisances such as smoky chimneys and, by means of clever contrivances, to make the ordinary business of living less tedious and toilsome. He lent, for a time, a word to the English language. Coleridge speaks of having his cottage chimneys at Nether Stowey rumfordized. That Gillray caricatured and Peter Pindar lampooned him shows how well known Rumford was in London society. In Gillray's cartoon he has a delightfully comical face. He stands with his back to one of his own stoves enjoying the warmth from it. He also enjoyed the cartoon, for he bought many proof impressions of *The Comforts of the Rumford Stove*, and gave them to his intimate friends observing, 'This is so much more like the Count than he is himself, that when you look upon it you cannot fail to think of your humble servant.' Peter Pindar in his wordy way is much less witty than his victim:

> Knight of the dish-clout, whereso'er I walk,
> I hear thee, Rumford, all the kitchen talk:
> Note of melodious cadence on the ear,
> Loud echoes 'Rumford' here and 'Rumford' there.
> Lo, every parlour, drawing-room, I see,
> Boasts of thy stoves, and talks of nought but thee.

Rumford was not the Count's original name. He had been Benjamin Thompson. It is an extraordinary tale, this tale of the store-keeper's apprentice, of North Woburn, Massachusetts, who read, and observed, and tested things for himself; attended courses on Experimental Philsophy at Harvard College; became a schoolmaster at Rumford (re-named Concord), married the squire's wealthy widow, took the side of the Crown in the American War of Independence, and escaped to England. In London he associated himself with Lord

George Germain. For a time he acted as one of the under-secretaries of State; then he went to the Continent, returned, was knighted by George III, and took service with the Elector of Bavaria. It was in Munich that he not only devised his classic experiments but made his name as a social theorist and organizer. His reforms in army and social administration led the Elector of Bavaria to make him, in 1792, a Count of the Holy Roman Empire.

In London the Institution associated with his name grew out of the *Society for Bettering the Condition and Increasing the Comforts of the Poor* founded by Thomas Barnard, with the Bishop of Durham, William Wilberforce and others in 1796.[1] In 1799 Barnard, at the suggestion of Rumford, set on foot the Institution first called the Rumford Institution and later, on receiving the King's Charter (January 13th, 1800), named *The Royal Institution of Great Britain*. Rumford, in a letter to Sir Joseph Banks, who was his friend, said that the Institution was equally his child.

Like the Pneumatic Institution in Bristol, it was established by private subscription. There was no institution expressly and originally founded for scientific research and scientific teaching until the foundation of the Royal School of Mines in 1851. It was Davy who made of the Royal Institution the first research institute, for he not only by his genius saved the whole enterprise from foundering but he changed its direction. It was originally established 'for diffusing the knowledge and facilitating the general and speedy introduction of new and useful mechanical inventions and improvements, and also for teaching, by regular courses of philosophical lectures and experiments, the application of these discoveries in science to the improvement of arts and manufactures, and in facilitating the means of procuring the comforts and conveniences of life'. Dr Bence Jones, historian of the Royal Institution, writes,[2] 'It had its origin in the work which Count Rumford did for the poor in Munich, and its primary objects were models, workshops, and useful knowledge to benefit the poor; lectures, researches and scientific experiments to amuse and interest the rich and to advance science, were comparatively the second intentions of the Founder.'

When Davy first joined the Institution it had already been moved

---

[1] Coleridge's poem addressed to Rumford was of this date.

[2] *The Royal Institution: Its Founder and its First Professors*, by H. Bence Jones, 1871.

from the house of Sir Joseph Banks, President of the Royal Society, to the beautiful house in Albemarle Street which remains, enhanced by time, quiet in its elegant proportions, in possession of itself. Before the house was bought for the Royal Institution in 1800 it was the home of Mr Mellish. Attics and laboratories were added, and Rumford himself designed the lecture theatre – once a ballroom.

Davy was first appointed as Assistant Lecturer in Chemistry, Director of the Laboratory, and Assistant Editor of the Journals of the Institution. He was to occupy a room in the house in Albemarle Street, be furnished with coals and candles, and be paid a salary of a hundred guineas a year. This appointment was made on February 16th, 1801; by March 11th the room had been prepared and furnished, and Davy had arrived in London and taken possession. The only thing he added to the furniture of the room was a little porcelain Venus given him by Thomas Wedgwood.

Six weeks after his arrival he gave his first short course of lectures on the galvanic phenomena which Luigi Galvani had first described in 1792. These lectures were noticed in *The Philosophical Magazine* (no. xxxv, p. 281), under the heading, *Royal Institution of Great Britain*:

> We have also to notice a course of lectures just commenced at the Institution, on a new branch of philosophy; we mean the galvanic phenomena: on this interesting branch Mr Davy (late of Bristol), gave the first lecture on the 25th of April. He began with the history of galvanism, detailed the successive discoveries, and described the different methods of accumulating galvanic influence . . . Sir Joseph Banks, Count Rumford, and other distinguished philosophers were present. The audience were highly gratified, and testified their satisfaction by general applause. Mr Davy, who seems to be very young, acquitted himself admirably well. From the sparkling intelligence of his eye, his animated manner, and the *tout ensemble*, we have no doubt of his attaining distinguished excellence.

Demonstrations of the effects of breathing nitrous oxide must also have taken place in Albemarle Street at this time, for Gillray's cartoon, *Scientific Researches! – New Discoveries in PNEUMATICKS! – or – an Experimental lecture on the Powers of Air* (Plate 3, p. 80), was drawn while Dr Garnett was still lecturer in chemistry and Davy his assistant. In the cartoon Davy 'seemeth elvish by his countenance'. His expression, as Gillray guyed it, is one of frolic anticipation of the effects of exploiting the powers of the 'air'. An interesting indication of the various figures in the cartoon is in a book of prints entitled *Account of the Caricatures of James Gillray*, ed. Thomas Wright and R. H. Evans, 1851.

When Dr Garnett, who had been in ill-health, and who had had many disagreements with Rumford, resigned his post, Davy was promoted from being assistant lecturer to the lectureship in chemistry. Dr Thomas Young was engaged as Professor of Natural Philosophy, Editor of the *Journal*, and general superintendent of the house. A sentence or two in one of Davy's letters to his former colleague at Clifton, the physiologist, Dr King, written soon after Dr Garnett had resigned and Davy had become sole lecturer in chemistry, shows one of the attractions of the Royal Institution as compared with the work at Clifton. He wrote, 'I have been nobly treated by the managers. God bless us! I am about a million times as much a being of my own volition as at Bristol. My time is even too much at my own disposal. So much for egotism, for weak, glorious, pitiful, sublime, conceited egotism.' Coleridge saw danger in this freedom. When he wrote to congratulate Godwin on Davy's settlement in London he said, 'I hope that his enchanting manners will not draw too many Idlers round him, to harass and vex his mornings.'

The Royal Institution precisely needed a young man of genius and eloquence who, like Davy, was not a political fanatic, though he numbered republicans among his friends. These included Thomas Richard Underwood, an acquaintance of Coleridge's, an artist of some talent with a fondness for science, a friend of Fuseli's, an antiquary, but ardent for the latest fashion in politics and morals; a flaming democrat and an admirer of Napoleon. He played a curious part in suggesting to Davy's earliest biographer stories about him which, though not exactly untrue, were coloured in their manner of telling by envy of a too-successful former associate. Underwood amused Coleridge. There is a letter from Coleridge to Davy dated May 20th, 1801, a reply to a request from Davy on behalf of Underwood for some information as to the character of a certain lady whose name Davy was not sure of – Hays or Taylor. Underwood thought of making her his mistress, but was cautiously seeking information first. Coleridge made great game of this. 'I rested my whole weight on my crutch, and laughed so that

[79]

I could scarce hold myself on the crutch, at the question, you put to me, in Underwood's name. I suppose, that when I had begun to laugh, from my exceeding weakness I continued it nolens volens.' What amused him so much was Davy's 'Hays *or* Taylor'.

It was with Underwood, or *Subligno*, as Coleridge sometimes called him, that Davy proposed an excursion to Penzance in the July following his London appointment. He had been instructed by the Managers of the Royal Institution, and encouraged by Sir Joseph Banks, to prepare a Course of Lectures on the Chemical Principles of the Art of Tanning to begin on the following November 2nd. Respectable persons of the trade – such respectable persons as were recommended by the Proprietors of the Institution – were to be admitted free. Mr Davy was given permission to absent himself during the months of July, August, and September for the purpose of making himself more particularly acquainted with the practical part of the business of tanning. Mr Davy, of course, knew one practical tanner pretty well already. Who could better aid him than Tom Poole? But first he would go home to Cornwall, looking in at Bristol on his way down. In July he wrote to Underwood in his most buoyant mood:

My dear Underwood,
   That part of Almighty God which resides in the rocks and woods, in the blue and tranquil sea, in the clouds and moonbeams of the sky, is calling upon thee with a loud voice: religiously obey its commands and come and worship with me on the ancient altars of Cornwall.

   I shall leave Bristol on Thursday next, possibly before, so that by this day week I shall probably be in Penzance. Ten days or a fortnight after, I shall expect to see you, and to rejoice with you.

   We will admire together the wonders of God, – rocks and sea, dead hills and living hills covered with verdure. Amen.

   Write to me immediately, and say when you will come. Direct H. Davy, Penzance. Farewell, Being of Energy!
                    Yours with unfeigned affection,
                                        H. DAVY.

When J. A. Paris was writing his biography of Davy, Underwood sent him this letter, together with an extract from his own journal kept at that time describing his visit. It is the most amusing account we have of Humphry Davy in Cornwall. Poor Underwood, that 'Being of

'Scientific Researches! New Discoveries in PNEUMATICKS! – or – an Experimental Lecture on the Powers of Air.' A cartoon by James Gillray. (See p. 78. Davy is holding the bellows)

Sep.r B.

...s much wanted in the
Laboratory of the Royal Institution.

Cleanliness

Neatness

Regularity. —

— The laboratory must be cleaned
every morning ... operations are
going on before ... o'Clock. —

— It is the business of Wm.
Payne to do this & it is
the duty of Mr. Davy to
see that it is done ... Take
care of & keep in order ...

— There must be in the laboratory.
Pen, Ink & paper & wafers &
these must not be kept in the
slovenly manner in which they usually
are kept. I am now writing with
a pen & ink such as was never used
in any other place. —

... are wanting, such
graduated glass tubes. Glass ...
measured to ... grains of mercury,

A Note in Davy's hand, in the Journal of the Royal Institution, 1809.

(See p. 137)

Energy', was walked off his legs. 'Trudged' is a significant word, and evidently Davy expatiated to such an extent on the beauties of Kynance Cove that Underwood was bored before he got there. Underwood wrote in his journal:

> On the 25th [of July], I went to Bristol, and on the 30th I arrived at Mrs Davy's at Penzance. On the 1st of August we set off on a pedestrian excursion, and proceeded along the edge of the cliffs, round the Lands End, Cape Cornwall, Saint Just, and Saint Ives, to Redruth and thence back to Penzance.

> Two days later we again started, and trudged along the shore to the Lizard. Kynance Cove had from the commencement of our intimacy been the daily theme of our conversation. No epithets were too forcible to express his admiration at the beauty of the spot; the enthusiastic delight with which he dwelt on the description of the serpentine rocks, polished by the waves and reflecting the brightest tints from their surfaces, seemed inexhaustible, and when we arrived at the spot he seemed absolutely entranced.

> During these excursions his conversation was most romantic and poetical. His views of Nature and her sublime operations, were expressed without reserve, as they were rapidly presented to his imagination: they were the ravings of genius; but even his nonsense was that of a superior being.

Underwood is always a little supercilious when writing of Davy, but we must be grateful to him for giving Paris the account of Davy and the irate landlady of the inn at Mullion. The story of this encounter is re-told by Paris:

> At the village of Mullion, a little incident occurred, which evinced the existence of that gastronomic propensity which, in after years, displayed itself in a wider range of operations. The tourists had, on their road, purchased a fine large bass of a fisherman, with the intention of desiring the landlady to dress it. On arriving at the inn, Mr Underwood retired to his room for the purpose of making some notes in the journal which he regularly kept. Davy had disappeared. In the course of a few minutes a most tremendous uproar was heard in the kitchen, and the indignant vociferations of the hostess, which, even with all the advantages of Cornish recitative, was not of the most melodious description, became fearfully audible.

G

Davy, it seems, had volunteered his assistance in making the sauce and stuffing for the aforesaid bass; and had he not speedily retreated, his services would have been rewarded, not by the scientific practice of appending a string of letters to his name, but in conformity with the equally ancient custom of attaching a certain dishonourable addition to the skirts of his jacket.

On his way back to London, Davy spent three weeks partly with the Beddoes and partly at Nether Stowey with Tom Poole, from whom he learnt much of the art of tanning. Poole was always struck by the speed at which Davy's mind worked, the quickness and truth of his apprehension. He said it was a power of reasoning so rapid that Davy himself could hardly be conscious of the process; it appeared to Poole as though it was pure intuition. In addition to the help Poole himself could provide, Davy had the aid of Poole's fellow-manufacturer, Samuel Purkis of Brentwood, to whom Poole introduced him, one who loved Davy living and honoured his memory when he was dead. The long and careful series of experiments initiated by Davy were mainly carried out at Brentwood, and went on long after the lectures had been delivered. It was said that Davy's interest in the pursuit could hardly have been keener if he had made it his trade. The true child of his grandmother, he never made the mistake of despising what had been learnt by skilled operators by rule of thumb or by tradition; he was attentive to processes governed by superstition. He initiated and superintended experiments, but he also questioned men. To his mother he wrote of a Penzance tanner who came to London at this time, 'I saw Mr William Bolitho and his brother-in-law yesterday, and they breakfast with me tomorrow. We are all fellows of the same craft; they are great practical tanners, and I am a theoretical one.' As a sportsman he had a personal interest in good leather and wore with much satisfaction two pairs of boots presented to him, one made of leather tanned by oak-bark, in the old way, and another by catechu. The catechu leather, the first that had ever been made, proved, Davy thought, not inferior in quality to the oak-bark leather.

That to be of use in the arts of life and in the management of heat was still very much the aim of the Royal Institution is clear not only from the lectures on tanning but also from letters from Davy to Gilbert in which Captain Trevithick's boilers are mentioned; and a letter from Trevithick to Gilbert in which Davy's friendly reception of him and Andrew Vivian is related. Gilbert, always a supporter of Trevithick, had urged Davy to bring the boiler to the notice of Count Rumford. Davy replied in November 1801 that he had not yet had an opportunity to submit the boiler, but would do so when the Count returned from a two months' visit to the Continent. Later he writes to Gilbert that he hopes to hear soon that the roads of England are the haunts of Captain Trevithick's dragons. But the very nature of the Royal Institution was to be changed by Davy's unexampled success as a lecturer, and by the original discoveries which brought fame, in almost equal measure, to himself and the Royal Institution between 1802 and 1812.

7 · *The Happy Orator*

Success as a lecturer came first. The lectures on the process of tanning and the chemical agents employed in it were delivered in the winter of 1801. Then, on January 21st, 1802, Davy delivered his *Discourse Introductory to a Course of Lectures on Chemistry*. This lecture, not originally intended for the press, excited so much interest that its publication was requested by part of the audience. It was printed in the April following its delivery, and is reprinted in the second volume of the collected *Works*.

Reading a lecture can never have the same effect on the mind as listening to it; yet, in reading it today, one is struck by the persuasiveness of an oration in no way declamatory; by the confident march of the thought from section to section; by the cohesion of the whole; by Davy's determined search for precision in the use of words, and by his temperate yet firm and vigorous statement of the new hope – the suitability of scientific studies to the progressiveness of man's nature. His authentic voice is heard in, 'And who would not be ambitious of becoming acquainted with the most profound secrets of nature, of ascertaining her hidden operations, and of exhibiting to men the system of knowledge which relates so intimately to their own physical and moral constitution.'

From a brief account of the recent advances made in human knowledge, he passes to the future, showing the present to be but the morning of a bright day yet to come. He indicates the immense field of research; the importance of the new agencies upon the improvement of society, and the fitness of scientific pursuits to bring delight to individuals. In speaking of society he has perhaps in mind memories of Ulysses' speech on degree in *Troilus and Cressida*; and of Hobbes's *Leviathan*, that commonwealth which Hobbes saw by analogy as an artificial man, with an artificial soul in the sovereign. For this sovereign soul Davy substitutes a common aim, a common search, uniting all ranks of society, each class contributing effectually to common support, the man of science and the manufacturer becoming more nearly assimilated to each other; new processes bringing about a diminution of labour; the rich and privileged orders becoming the guardians of civilization and refinement, the friends and protectors of the labouring part of the community. He says, 'The unequal division of property and of labour, the difference of rank and condition amongst mankind, are the sources of power in civilized life, its moving causes, and even its very soul; and in considering and hoping that the human species is capable of becoming more enlightened and more happy, we can only expect that the great whole of society should be ultimately connected together by means of knowledge and the useful arts; that they should act as the children of one great parent, with one determinate end, so that no power may be rendered useless, no exertion thrown away. In this view we do not look to distant ages, or amuse ourselves with brilliant, though delusive dreams concerning the infinite improveability of man, the annihilation of labour, disease and even death. But we reason by analogy from simple facts. We consider only a state of human progression arising out of its present condition. We look for a time that we may reasonably expect, for a bright day of which we already behold the dawn.'

Davy had been born during the American War of Independence; had been a boy when, in Blake's words, 'Shadows of Prophecy shiver along the lakes and the rivers, and mutter across the ocean: "France, rend down thy dungeon."' His late boyhood and early manhood had been marked by war. When he delivered his lecture at the beginning of 1802 there was prospect of peace; the Treaty of Amiens was formally signed in March of that year. Not only revolutionaries like Underwood but moderate 'liberals', such as Tom Poole, were, with Fox and a host of the English, to visit France. It was no wonder that Davy should preach not tension but co-operation; not anarchy but order; not mad haste but slow progression. His upbringing had made him conscious that a monopoly of virtue or of wickedness was not held by any one class. Of manufacturers he knew best Poole and Purkis, good men; among the great landowners with whom he was to be associated on the Board of Agriculture were those mindful of the public good. He was young enough to be optimistic; he was of a new generation weary of warfare. His words still convey the earnestness of an impassioned speaker.

It is to Purkis that we are indebted for some of the most vivid descriptions of Davy's personal success as a lecturer, success which began in 1802 and continued until he resigned his Professorship at the time of his marriage in April 1812. He was made Professor of Chemistry in May 1802, and elected Fellow of the Royal Society in 1803.

Purkis wrote of Davy's success during his first course of lectures on Chemistry in 1802:

> The sensation created by his first course of Lectures at the Institution, and the enthusiastic admiration which they obtained, is at this period [Purkis was writing to Ayrton Paris after Davy's death] scarcely to be imagined. Men of the first rank and talent, – the literary and the scientific, the practical and the theoretical, bluestockings, and women of fashion, the old and the young, all crowded – eagerly crowded the lecture-room. His youth, his simplicity, his natural eloquence, his chemical knowledge, his happy illustrations and well-conducted experiments, excited universal attention and unbounded applause. Compliments, invitations and presents, were showered upon him in abundance from all quarters; his society was coveted by all, and all appeared proud of his acquaintance.

The Royal Institution became not only the fashion but the rage. Count Rumford wrote to his daughter Sally that the 'nice able man' they had found for the Institution was drawing 'crowds of the first people, that it was certainly gratifying to see the honourable list of lords, dukes, etc., as fifty-guinea subscribers'.

The ladies praised the lecturer's bright eyes and said they were meant for something other than poring over crucibles; they sent him notes and sonnets. His eloquence stirred them; they enjoyed the experiments and the illusory feeling of knowledge the lectures gave them. Nor were the feelings altogether illusory when it was Davy who was lecturing. He had a quick command of language as well as a sparkling eye. He was persuasive. If fortune had not directed his energies into the channel which made him one of the greatest chemists in Europe she would have made him not a poet but a converting preacher – converting to science. He had what we in Cornwall call 'the gift' for which, if not possessed, there is no substitute in preaching. At one time he was pressed to take orders in the Church of England.

There were dangers in a fashionable audience. Francis Horner wrote in his journal for March 31, 1802:

The Happy Orator

I have been once to the Royal Institution and heard Davy lecture on animal substances to a mixed and large assembly of both sexes, to the number perhaps of three hundred or more. It is a curious scene: the reflections it excites are of an ambiguous nature; for the prospect of possible good is mingled with the observation of much actual folly. The audience is assembled by the influence of fashion merely; and fashion and chemistry form a very incongruous union. At the same time it is a trophy to the sciences; one great advance is made towards the association of female with masculine minds in the pursuit of knowledge, and another domain of pleasing and liberal inquiry is included within the range of polished conversation. Davy's style of lecturing is much in favour of himself, though not perhaps entirely suited to the place; it has rather a little awkwardness, but it is that air which bespeaks real modesty and good sense: he is only a little awkward because he cannot condescend to assume that theatrical quackery of manner which might have a more imposing effect. This was my impression from his lecture. I have since [April 2nd] met Davy in company, and was much pleased with him; a great softness and propriety of manner which might be cultivated into elegance; his physiognomy struck me as being superior to what the science of chemistry in its present plan can afford exercise for; I fancied to discover in it the lineaments of poetical feeling.[1]

In 1809 this same Scotsman – he was at one time Member for the borough of St Mawes and later for St Ives – wrote in a letter to Lord Webb Seymour:

I hear faint and distant rumours of immense discoveries in chemistry which I eagerly wish I had the means of knowing and following; but it has pleased the Gods to dispose of me otherwise, for no good to others, and for less enjoyment to myself. I hear, however, of Davy, Berzelius, and others, from Tennant occasionally; and their late successes appear truly wonderful and immensely important. How happens it that Edinburgh contributes nothing to these discoveries, with all the study and zeal that prevail there for this science?

Even the earlier audiences were not as frivolous as the sober Scot imagined. John Dalton came from Manchester to lecture in the Royal

---

[1] *Memoirs and Correspondence of Francis Horner*, ed. Leonard Horner, 1843, p. 109.

Institution in December 1803.[1] He lectured not merely on chemistry but on mechanics and physics, repeating the course in 1809. Of his audiences in 1803 he wrote to his brother:

> The number attending were from one to three hundred of both sexes, usually more than half men. I was agreeably disappointed to find so learned and attentive an audience, though many of them of rank. It required great labour on my part to get acquainted with the apparatus and to draw up an order of experiments and repeat them in the intervals between the lectures, though I had one pretty expert to help me . . . The scientific part of the audience was wonderfully taken with some of my original notices relative to heat, the gases, etc., some of which had not been published. Had my hearers been generally of the description I had apprehended the most interesting lectures I had to give would have been the least relished; but, as it happened, the expectations formed had drawn several gentlemen of first-rate talents together; . . .

This letter was written in December 1803; on January 10th Dalton wrote another letter in which he gives one of the best pictures we have of the young Davy's way of preparing his early lectures. It is amusing to find Dalton, the Quaker, in whom one would have expected some austerity, complaining that Davy's principal failing was that he did not smoke. The relations between the younger and older man – Davy was twenty-five and Dalton thirty-eight at this time – seem easy and cordial. Davy's subsequent relations with Dalton, his attitude to the Atomic Theory and to the doctrine of Definite Proportions will be referred to in later chapters. Of his experience with Davy in his early days at the Royal Institution Dalton wrote:

> I was introduced to Mr Davy, who has rooms adjoining mine in the Royal Institution. He is a very agreeable and intelligent young man, and we have interesting conversations in the evening. The principal failing in his character is that he does not smoke. Mr Davy advised me to labour my first lecture; he told me the people here would be inclined to form their opinion from it. Accordingly, I resolved to write my first lecture wholly, to do nothing but to tell them what I would do and enlarge on the importance and utility of

---

[1] *John Dalton*, by Sir Henry E. Roscoe, 1901, pp. 163–6.

science. I studied and wrote for nearly two days, then calculated to a minute how long it would take me reading, endeavouring to make my discourse about fifty minutes. The evening before the lecture Davy and I went into the theatre. He made me read the whole of it, and he went into the farthest corner; then he read it and I was the audience. We criticised upon each other's method. Next day I read it to an audience of about 150 or 200 people, which was more than were expected. They gave a very general plaudit at the conclusion, and several came up to compliment me on the excellence of the introductory . . .

The very variety of the work of necessity undertaken by Davy during the first six years of his residence at the Royal Institution was made by him to serve his main object – original research in galvanism and electrochemistry. There was nothing neat and small in his aims. Just as in the making of a great poem or play there must be an immense multiplicity contained within the controlling oneness of design and effect, so with Davy in the conduct of his quest for fresh knowledge. He could wait, attending to other affairs, but always with the hope of using them to light up his main design to which his most intense application was directed.

He multiplied his observations in the world, experimented in the laboratory, submitted his results to thought, tabulated, lectured, demonstrated, and published. The facts he had ascertained relating to tanning were communicated to the Royal Society in 1803 and published in the *Philosophical Transactions*. In 1805 he gave his first lectures in geology, a science then in its infancy; and he made these early lectures the basis of a Course he delivered in 1811. We have, too, his Introduction to a projected series of papers never completed, his *Sketches of the Geology and Mineralogy of Cornwall*, that 'county of veins' in which, with Gregory Watt, he had made his earliest explorations. Some of his best descriptive writing and excellent examples of his eloquence are to be found in his geological lectures and observations. He never wearied of pointing out that mineralogy ought to be a preparation for geology, and considered as affording the characters by which its mysteries are deciphered; and that it is in the great arrangements of nature, and not in the details of the museum, that the facts and foundations of the science must be sought for and examined. In the midst of his ordered expositions of existing and conjectured knowledge he celebrates the grandeur and endurance of granite or the splendour of serpentine; he recalls his sensations on the short green grass, tufted with heath and furze, which cover sparsely the cliffs above Trerine Downs in west Cornwall; or remembers the hours spent resting on a rock in Scotland's valley of the Awe. It is in the introductory lectures to his own Courses, or in Introductions to the lectures generally which were to be delivered during the Session at the Royal Institution, that his gifts as an orator find best scope. But sometimes, too, in the midst of some demonstration, or in the exhibition of a drawing to illustrate his subject-matter, his mind is kindled. In a geological lecture, after exhibiting a painting of the granite cliffs of the Land's End, he expatiates on the texture of granite, one of the firmest of stones, and on its suitability to the structure of enduring edifices and lasting memorials. No wonder the sculptor Chantrey liked to talk with Davy and Wollaston, and enjoyed the study of geology.

Comparatively few of the large number of popular lectures delivered by Davy were preserved. They were, his brother says, little regarded by Humphry once they had served the special purpose for which they were written. But a distinct body of work remains from his agricultural lectures. In 1802 he was first invited by Arthur Young, Secretary to the Board of Agriculture, to give a course of lectures to the members on the connexion of chemistry with vegetable physiology. It was part of the wide effort to convert farming 'from a mere art of blind processes into a rational system of science'. For ten years successively Davy delivered lectures to the Board of Agriculture at its meetings, lectures which were published in 1813 under the title, *Elements of Agricultural Chemistry.*

Davy's arduous work in association with agriculture was also part of his pleasure. The companionship and hospitality of great landowners which he began to enjoy gratified not only any social ambitions he may have had but administered to an essential want of his nature. With the hospitality of the country house went often the hospitality of fishing rights. He was so passionate an angler that he could cheer himself up in Albemarle Street by merely looking at his artificial flies. He was half a farmer's son, and his father, as John Davy points out, being of a speculative turn of mind, had not confined himself to common routine methods but had experimented. In the quietness and solitude of Somerset Davy had talked, too, with Poole, a farmer as well as a tanner. An interest in the application of science to agriculture was not adventitious but deep-rooted in him. Through his connexion with Arthur Young, Secretary to the Board, he became associated with the most progressive agriculturalists of the day.

In addition to his routine work as lecturer at the Royal Institution it fell to Davy to pronounce the eulogium of the three of his predecessors he most admired in Science. Henry Cavendish died in 1810. Of him Davy said in the course of a chemical lecture:

Of all the philosophers of the present age, Mr Cavendish was the one who combined, in the highest degree, a depth and extent of mathematical knowledge with delicacy and precision in the methods of experimental research. It may be said of him, what can, perhaps, hardly be said of any other person, that whatever he has done has been perfect at the moment of its production. His processes were all of a finished nature. Executed by the hand of a master, they required no correction; and though many of them were performed in the very infancy of chemical philosophy, yet their accuracy and their beauty have remained amidst the progress of discovery, and their merits have been illustrated by discussion, and exalted by time.

In general, the most common motives which induce men to study are, the love of distinction, of glory, or the desire of power; and we have no right to object to motives of this kind; but it ought to be mentioned, in estimating the character of Mr Cavendish, that *his* grand stimuli to exertion were evidently the love of truth and of knowledge. Unambitious, unassuming, it was with difficulty that he was persuaded to bring forward his important discoveries. He disliked notoriety, and he was, as it were, fearful of the voice of fame. His labours are recorded with the greatest dignity and simplicity, and in the fewest possible words, without parade or apology; and it seemed as if in publication he was performing, not what was a duty to himself, but what was a duty to the public. His life was devoted to science, and his social hours were passed amongst a few friends, principally members of the Royal Society. He was reserved to strangers, but, when he was familiar his conversation was lively and full of varied information. Upon all subjects of science he was luminous and profound; and in discussion wonderfully acute. Even to the very last week of his life, when he was nearly seventy-nine, he retained his activity of body, and all his energy and sagacity of intellect. He was warmly interested in all new subjects of science; and several times in the course of the last year witnessed, or assisted in some experiments which were carried on in this theatre, or in the laboratory below.

Since the death of Newton, if I may be permitted to give an opinion, England has sustained no scientific loss so great as that of Cavendish. Like his great predecessor, he died full of years and of glory. His name will be an object of more veneration in future ages than at the present moment. Though it was unknown in the busy scenes of life, or in the popular discussions of the day, it will remain illustrious in the annals of science, which are as imperishable as that nature to which they belong; and it will be an immortal honour to his house, to his age, and to his country.

8 · Discovery

It was in the year 1807 that Davy made the discovery for which he is best remembered in the history of science – the isolation of the elements potassium and sodium.

He had brought all that was enlightening in the varied excursions of his mind to vivify his work on electrochemistry. Before leaving Bristol he had experimented with the voltaic pile possessed by Beddoes. His first lecture at the Royal Institution was on Galvanic Phenomena. By 1806 he was ready with an important contribution to thought. It was communicated in his first Bakerian Lecture, *On Some Chemical Agencies of Electricity*, read before the Royal Society on November 20th, 1806, and published in the *Philosophical Transactions* for 1807.

In his 'Historical View of the Progress of Chemistry' (*Elements of Chemical Philosophy*, 1812) Davy emphasizes the fact that it was not till the era of Volta's wonderful discovery in 1800 of a new electrical apparatus that any great progress was made in chemical investigation by means of electrical combinations. 'Nothing', he said, 'tends so much to the advancement of knowledge as the application of a new instrument. The native intellectual powers of men in different times are not so much the causes of the success of their labours as the peculiar nature of the means and artificial resources in their possession. Without the voltaic apparatus there was no possibility of examining the relations of electric polarities to chemical attractions.'

His Bakerian Lecture of 1806 is an example of an inquiry made upon the hidden sensible properties of things and upon existing relations of facts. It proved that, in addition to being a brilliant experimenter, he had what Coleridge called an *arranging* head; it made voltaic electricity in connexion with chemistry, and as a new power of chemical analysis, a subject of intense interest to his peers; it won him the kind of praise from a learned audience which his lecture of 1802 had won from a popular one; it laid the foundation of his European

[93]

reputation as a scientist – Berzelius considered it one of the most remarkable of all contributions to the theory of chemistry. A year after its publication Davy was awarded by the French *Institut* the prize of 3,000 francs, founded by Napoleon when first Consul, for the most important results in electrical research during each year. Even Charles Lamb, who used to say that in everything that belonged to science he was a whole encyclopedia behind the rest of the world, wrote to his friend Manning a lighthearted comment on this award made though the two countries were at war. He said that Napoleon had voted 5,000 livres to the great young English chemist, but that it hadn't arrived yet.

Towards the close of his paper of 1806 Davy stated the facts which induced him to hope that the new mode of analysis might lead to the discovery of the *true* elements of bodies; and he rounded off the whole lecture with the words: 'Natural electricity has hitherto been little investigated, except in the case of its evident and powerful concentrations in the atmosphere. Its slow and silent operations in every part of the surface will probably be found more immediately, and importantly connected with the order and economy of nature; and investigations on this subject can hardly fail to enlighten our philosophical systems of the earth; and may possibly place new powers within our reach.' For the rest of his life his imagination was haunted by those 'slow and silent operations'; their allure is the true clue to his passionate wanderings; he was like a magician who had glimpsed some astounding simplicity, which if only he could remember and apply would unlock concealed secrets beyond all dreams.

But first he turned to the laboratory use of electrolysis. In October 1807, ten months after the hope stated in his paper of 1806, he had, by the decomposing action of an electric current, proved first that potash and then that soda were not elements but oxides of metals. As assistant in the laboratory at that time Humphry had a cousin, Edmund Davy, whom he had introduced to the service of the Institution, and who afterwards became Professor of Chemistry to the Dublin Society. To him we owe the description of Davy's delight when he saw the minute shining globules burst through the crust of the potash and take fire as they entered the atmosphere. Humphry danced for joy round the room, and in his laboratory notebook he wrote in large letters 'CAPITAL EXPERIMENT, PROVING THE DECOMPOSITION OF POTASH'. Some days later the metallic base of soda was disclosed. Davy named the substances which he had isolated potassium and sodium, and while his mind was still hot from the discovery, he composed his second Bakerian Lecture which was delivered on November 19th, 1807, winning him more fame than even his restless imagination could have anticipated in Penzance. In the second section, after describing the methods he had used, he wrote:

... When the power of 250 was used, with a very high charge for the decomposition of soda, the globules often burnt at the moment of their formation, and sometimes violently exploded and separated into smaller globules, which flew with great velocity through the air in a state of vivid combustion, producing a beautiful effect of continued jets of fire.

In the fourth section of his lecture he noted that after he had detected the 'bases' of the fixed alkalies he had found it difficult to preserve and confine them so as to examine their properties and submit them to experiments; 'for, like the *alkahests* imagined by the alchemists, they acted more or less on every body to which they were exposed'. The best prison he found for them was recently distilled naphtha. In language at once vivid and precise he described the experiments made to determine their properties and nature. He classed the new soft, lustrous, light substances as metals, and said he had ventured to name them potassium and sodium, though he said the words were more significant than elegant. He wanted words which would imply simply the metals produced from potash and soda, words avoiding any theoretical expression for, he said, 'the mature time for a complete generalization of chemical facts is yet far distant'.

He had not, of course, any foreknowledge of the natural transmutation of the radioactive elements; he had not achieved a real transmutation such as that of which the alchemists dreamed. But in the thirteenth century he might have been thought a wizard. Lustrous metals that would float, plucked from potash and soda, would have seemed the result of a most powerful magic. The changeableness of changeable things had never before been so vividly demonstrated. The new products themselves could be furiously active devils. Davy meant to use them as agents in further enterprises: 'In themselves they will undoubtedly prove powerful agents for analysis; and having an affinity for oxygen stronger than any other known substances, they may possibly supersede the application of electricity to some of the undecompounded bodies.' He had in mind to wreck the earths as he had wrecked the fixed alkalis.

Instead he was very nearly wrecked himself. In the December of 1807, when he was at the height of his popularity, when his lectures were thronged, when ladies wrote him poems and all men invited him, he fell ill and came so near death that he was unable to resume work for many weeks. By some it was thought that the illness was caused by the fumes from baryta, an alkaline earth of great weight on which he had been working; and various forms of the following squib were passed round:

> Says Davy to Baryt, 'I've a strong inclination
> To try to effect your deoxidation;'
> But Baryt replies – 'Have a care of your mirth,
> Lest I should retaliate, and change *you* to earth.'

It was also suggested that the illness was caused by Davy's breathing the foul air of Newgate while working at a scheme to improve the prison's ventilation; but according to his friend Dr Babbington, one of the physicians attending him, the deadliness of the attack and the long-protracted weakness were due to over-fatigue and excitement. Great concern was shown at his illness. Bulletins were issued, and eulogies were pronounced of a kind not usually made until after a man is dead. With the housekeeper of the Institution Davy was a favourite. She proved an excellent nurse. Communication was constant with Penzance and, as he began to get better, it was of Penzance he thought, longing for, and receiving, apples from a tree he had planted as a boy, and various objects he fancied, including an old teapot. No doubt it made better tea than grander pots. He always looked back with quick affection to the life of his boyhood, though he could not return to it any more than he could return to the innocent faith of his childhood, though he considered a strong religious faith the happiest blessing that could be bestowed on a man. With a renewal of health he felt a resurgence of that passionate intuition of the immortality of the better part of man's nature which he was to try again and again to express; a return of buoyancy brought with it the feeling that once again he could relish versing, and affirm the unchanging in the changeable. He revised a poem he had begun many years before:

Discovery

Written after Recovery from
a Dangerous Illness

Lo! o'er the earth the kindling spirits pour
    The flames of life that bounteous Nature gives;
The limpid dew becomes the rosy flower,
    The insensate dust awakes, and moves, and lives.

All speaks of change: the renovated forms
    Of long-forgotten things arise again;
The light of suns, the breath of angry storms,
    The everlasting motions of the main.

These are but engines of the Eternal will,
    The One Intelligence, whose potent sway
Has ever acted and is acting still,
    Whilst stars, and worlds, and systems all obey;

Without whose power, the whole of mortal things
    Were dull, inert, an unharmonious band,
Silent as are the harp's untuned strings
    Without the touches of the poet's hand.

A sacred spark created by His breath,
    The immortal mind of man His image bears;
A spirit living 'midst the forms of death,
    Oppressed but not subdued by mortal cares!

A germ preparing in the winter's frost
    To rise, and bud, and blossom in the spring;
An unfledged eagle by the tempest tost,
    Unconscious of his future strength of wing.

The child of trial, to mortality
    And all its changeful influences given;
On the green earth decreed to move and die,
    And yet by such a fate prepared for heaven . . .

Life had not waited for him while he lay ill. His fellow philosophers were swift as hounds on the track he had indicated, and by the end of June 1808 he was ready himself with a paper on the decomposition of the alkaline earths. He described how Berzelius of Stockholm had succeeded in decomposing barytes and lime and how he, following the 'happy method' described to him in a letter from Berzelius, had separated metallic bases from barytes, strontia, lime, and magnesia, and proposed as names for the new elements, barium, strontium, calcium, and magnium, saying that the last of these words was undoubtedly objectionable, but magnesium had already been applied to metallic manganese, and consequently would be an equivocal term. Later he allowed magnesium. Had he procured the further metallic substances he was in search of he would have proposed for them, he said, the names of silicium, alumium, zirconium, and glucium. He did claim to have isolated alumium from alumina in 1812; this became, because of the greater harmony in sound with potassium and sodium, aluminium.[1] His work on the new elements and amalgams caused him to nourish various new theories as he observed fresh facts; but he repeatedly said that he did not attach much importance to these notions: 'The more subtle powers of matter are but just beginning to be considered, and all general views concerning them must as yet rest on but feeble and imperfect foundations.'

He became as it were the novelist of 'bodies', trying to compel them to give up their secrets so as to relate their case histories. He was especially concerned with oxymuriatic gas which he was to re-name chlorine from its greenish colour. Lavoisier thought it was the oxide of a radical; Davy having 'put it to the question' in a variety of ways decided, cautiously, that it might be presumed an element. He stated to the Royal Society the facts which inclined him to believe 'that the body, improperly called *oxymuriatic gas*, has not yet been decompounded, but that it is a peculiar substance, elementary as far as our knowledge extends, and analogous in many of its properties to oxygen gas'. In a lecture of 1811, referring to presumed elements, he said, 'In modern philosophy all views of known *ultimate indestructible* elements are discarded. Those forms of matter which have not yet been changed by art are considered as *undecompounded*, but this is merely with respect to the present state of knowledge. The experimental philosopher does not consider those substances which are elementary to him as necessarily elementary in the great operations of the universe. He merely follows nature in the great path of experiment. The number of prin-

[1] Davy contributed some thirty-five electrical and chemical terms to the English vocabulary. See *The Royal Society and English Vocabulary*, by A. D. Atkinson, *Notes and Records of the Royal Society of London*, vol. 12, no. 1, pp. 40–3.

ciples not decompounded is consequently in a state of change. Till very lately the earths, the alkalies, and certain acids were considered as elements; but all these it has been my good fortune to decompose. The arrangements of the science change with every new acquisition of facts.'

Davy's demonstration that the strongest agents could not elicit oxygen from chlorine not only allowed him to presume its elementary nature but also challenged Lavoisier's generalization that all acids must contain oxygen, a generalization which had hardened into a dogma. John Davy considered that, in relation to the facts and doctrines of science, this inquiry of his brother's into the nature of chlorine, and the further researches to which it led, broke down prejudices and, as Davy remarked in an early lecture on chlorine, 'in the physical sciences there are much greater obstacles in overcoming old errors than in discovering new truths; the mind being in the first case fettered, in the last perfectly free in its progress'.

John Davy worked under his brother in the Royal Institution from the winter of 1808 until 1811 before going on to take his medical degree in Edinburgh. He had gone to school at Helston and Barnstaple, and had been as eager to leave Barnstaple as Humphry had been to leave Truro. A letter written by Humphry to his mother is of interest in relation to Humphry's treatment of Faraday later. He told his mother that it would be illiberal in him, even if he had the power, to bring John into the Royal Institution on a salary for which he could do little or nothing to promote the interests of the establishment. 'All I can do with propriety is to get apartments assigned to him; coals, candles, attendance, and so on, he will have in my rooms.' He then continues, 'Now I live very little in the Royal Institution. I never dine there, and when I do not dine with some of my friends I dine at a coffee-house. It would be fatal to John's improvement, even if I had the power, to take him with me wherever I go; and for him to live exactly as I do before he has the means of getting what I get, would spoil him for an economist; and it will not be possible for me to alter my mode of living and to make myself a hermit on his account.' He proposed that John and his cousin Edmund should mess together when he was not there: 'Edmund will teach him economy, which is a very great virtue, and I will endeavour to teach him chemistry and philosophy.' He proposed that his mother should advance from time to time the sums that she had paid during John's last year at school, and that he should take care of the rest.

He was a little anxious always that his sisters should not grow discontented at Penzance. In 1806 Kitty had paid him a visit in London. His Aunt Sampson, 'always good and kind to me', also paid him a visit. While Kitty was with him he wrote Betsy a letter of praise mingled with brotherly admonition:

January 2, 1806.

My dear Sister,

I wish I had a flying wooden box, such a one as you have read of, or ought to have read of, in the 'Arabian Nights' Entertainments'; for then I would come and see you every morning, and take you an airing about the world, and show you all the great cities, and great sights, people long so much to visit, and so soon grow tired of when they are wise.

But you are happier than if you were a traveller; and, I am sure happier than if you were with Kitty in London; for you are contented, and living with old friends, whom, I am sure, you will always be wise enough to prefer to new ones. You do not know how much I was delighted with your conduct whilst I was at home. Your strong affection for my mother, your love for John, your attention to all your duties, gave me the warmest pleasure. I hope you will always be as good, and as unaffected, and as happy as you are now; and that you will enjoy all the blessings that Heaven can bestow on you.

I trust, my dear girl, that you will endeavour to improve yourself in writing, and that you will persevere till you can write and spell very correctly. I shall be very glad to have you for a correspondent; and I will answer every correct and well-written letter that you send me.

I enclose a one pound note, which you will lay out in books, or in anything else that you like. I enclose another one pound note, which I wish to have disposed of in the following manner, from me: – to Mary Launder, 5*s*.; to Betty White, 5*s*.; and with the rest you will buy some ribands, or little articles of dress for the Doctor's Jenny, my aunt Sampson's Phyllis, my aunt Millet's maid, and my mother's servant, as new year's gifts.

I am, dear Betsy,

Your affectionate Friend and Brother,

H. DAVY.

Discovery

At the summit of his fame he took trouble to get new gowns for his sisters. The pattern that had been given him was too dilapidated to be of any use and so he wrote asking for measurements. It is pleasant to find the distinguished lecturer and discoverer, sought after by all the learned and famous people, playing the part of the brother in Town to his sisters at Penzance. The dresses must have been a peculiar pleasure to them as coming from Humphry in London. He writes to his youngest sister:

> I looked last week at the pattern of the gown that my sister put into my hands, and found it so worn and tattered that nothing can be made of it; I cannot therefore get your gowns made till you send me another. The best way will be to give me measure of the waist, shoulders, length, etc., in this way, and there can then be no difficulties: thus waist, 15 inches, or whatever it may be; between shoulders: length from waist to skirt or train.
>
> I do not wish to send gowns you cannot wear, and in this way they can be well made. By a piece of tape you can easily measure and then try the length by a carpenter's rule, and give me the results for yourself, and for Kitty and Grace, and I shall then be able to send your gowns a few days after I receive your letter.[1]

He calls this 'as stupid a letter as ever was written about gowns'; but except that 15 inches seems an astonishingly small assessment for a waist, even in the days of astonishing waists, it is a most practical letter for a man to have written.

In his earlier days in London he did sometimes entertain in his room at the Royal Institution. Clement Carlyon, a Truro man, knew Davy intimately; he had also seen a good deal of Coleridge when both were in Germany. He had gone from Truro Grammar School to Pembroke College, Cambridge, where he became a Fellow; later he studied medicine in London, and, as a physician, settled in Truro. In his book *Early Years and Late Recollections* he has a description of Davy's being put to confusion at a little dinner-party in his own quarters at the Royal Institution. Carlyon writes:

> I remember a laughable circumstance, which occurred at a dinner which he gave to a few friends at his rooms in Albemarle Street. Three fine-looking woodcocks were put on the table as a

[1] Printed in *Humphry Davy: Poet and Philosopher*, T. F. Thorpe, 1896, p. 116.

remove – and the great philosophers present, Dr Young, Dr Babington, and one or two others, as well as the humble individual who records this anecdote, evidently anticipated a treat, in assisting to eat up our host's Cornish delicacies. With a little assistance in the carving department, the guests were all served, in no long time, and the attack commenced – but proceed we could not; all tasted, we knew not what; all looked, we knew not how; and almost simultaneously exclaimed, 'Why, what in the world have we here, Davy, in the shape of woodcocks?' In short these dainties were totally uneatable; and the confession made frankly by our friend was, that desirous of keeping the woodcocks, which he had received out of Cornwall three weeks previously, for our dinner-party, he had placed them in the exhausted receiver of an air-pump, which had previously been employed by him in making some experiments with ether. *Hinc nostrae lachrymae!* Tears, certainly of merriment, more than of disappointment, for the laugh against poor Davy (who protested that he could not have conceived the possibility of a more perfect vacuum than was effected by him, prior to depositing the birds in the receiver) went far to compensate us for our epicurean mishap.

Carlyon said he believed no other philosopher was ever capable of combining pleasure with the graver pursuits of science so intensely as Davy. His last tête-à-tête with him, before he himself went off to live in Truro, led to much lively talk. He writes:

My last tête-à-tête with Sir Humphry Davy occurred accidentally in the following manner. I was in town prior to my settling in Truro professionally, and went to a coffee-house in Leicester Square to dine, as I often had done very satisfactorily and economically, on macaroni, nowhere better prepared than there. I was scarcely seated when Sir Humphry entered the room, and we were much too old friends to dine apart. The macaroni was of course ignored, and I readily acquiesced in Sir Humphry's superior catering.

He was then in the habit of faring sumptuously every day, being most days the favoured guest of some person of rank and fortune. Nevertheless our dinner was sufficiently recherché, and our conversation did not flag for want of a glass of spritely wine. For that afternoon we gave philosophy a truce; and I well remember that –

Much of love
Of beauty much the sprightly discourse ran.

It was just then that Sir Humphry [not, however, then *Sir Humphry*] was at the height of popularity as a lecturer in Chemistry at the Royal Institution; and it appeared that he was constantly receiving complimentary notes, emanating much more from the bower of Cupid and the Muses, than from the studio of Newton or Lavoisier.

Davy made friends in various parts of the country as well as in London. When he was preparing his lectures on geology and agriculture his work and pleasure took him much out of town. In 1803 he was introduced by Sir Joseph Banks, who, like the King, was a progressive farmer on his great estate, to T. A. Knight, a pioneer in plant physiology. When Davy first visited him, Knight lived at Elton, near Ludlow; later he moved to Downton Castle on the Teme. Davy loved to be Knight's operator in experiments. He became the friend of the Knight family. Mrs Stackhouse Acton, Knight's daughter, relates how surprised they were, when Davy first visited them, that he should look so young and yet be already famous. She recalled his good nature to her and the other children, and, alluding to Davy's visits, said that her father, to the end of his life, spoke of them as affording many of the pleasantest days he ever passed. Davy, when he returned to London from Wales, would feel melancholy at the sight of flies sporting in the sunshine on a fine day and think, 'What such a day would be worth at Downton!' He told Knight in one of his letters that he was almost glad when the wind blew from the east for the first week of his return, for then he did not so much long for the Teme.

Another comment on Davy's youthful appearance, after he had spent six years in London, was made by Benjamin Haydon, the painter, who was originally of Plymouth. He describes a dinner-party at Sir George Beaumont's at which he himself, a happy and excited guest, had arrived before Davy came in. Haydon relates of Davy's entry, 'Mr Davy was announced, and a little slender youth came in, his hair combed over his forehead, speaking very dandily and drawlingly ... Davy took Lady Beaumont, the rest followed as they pleased.' Haydon then goes on to say, 'Davy was very entertaining, and I well remember a remark he made which turned out a singularly successful prophecy; he said, "Napoleon will certainly come in contact with Russia by pressing forward in Poland, and *there* probably will begin his destruction." '

9 · Coleridge Again: Ireland

Of the effect of Davy's Bakerian Lecture of 1807 on his audience, and of his subsequent illness, Coleridge wrote to Dorothy Wordsworth on November 24th of that year:

> I found Davy, this morning, in bed, very seriously unwell – and am going to sit with him this evening. He had exposed himself to too violent alternations of Cold and Heat during the March to Glory, which he has run for the last six weeks – within which time by the aid and application of his own great discovery, of the identity of electricity and chemical attractions, he has placed all the elements and all their inanimate combinations in the power of man . . . I was told by a fellow of the *Royal Society*, that the sensation produced last week by the reading of his Paper there, was more like stupor than admiration – and the more, as the whole train of these discoveries have been the result of profound Reasoning, and in no wise of lucky accident. This account will probably interest William . . .

Coleridge's vivid imagination leapt beyond Davy to ultimate possibilities and he uttered, as so often, his warning of the danger if the source of virtue were disregarded:

> Davy supposes that there is only one power in the world of the senses; which in particles acts as chemical attractions, in specific masses as electricity, and on matter in general, as planetary gravitation. Jupiter est, quodcumque vides; when this has been proved, it will then only remain to resolve this into some Law of vital Intellect – and all human knowledge will be Science and Metaphysics the only Science. Yet after all, unless all this be identified with Virtue, as the ultimate and supreme Cause and Agent, all will be a worthless Dream. For all the Tenses and all the Compounds of Scire will do little for us, if they do not draw us closer to the *Esse* and *Agere*.

Coleridge Again: Ireland

Coleridge's grief when he knew that Davy was not only ill but his life precarious, his recovery doubtful, was expressed in a letter to Southey on December 14th. 'And to this day, no distinct symptom of Safety has appeared – tho', today he is better. I cannot express what I have suffered – Good heaven! in the very spring tide of his Honors – his? his Country's! the World's! after discoveries more intellectual, more ennobling and impowering human Nature, than Newton's! But he must not die.' When Coleridge wrote of himself in *The Tombless Epitaph*:

> Sickness, 'tis true,
> Whole years of weary days, besieged him close
> Even to the gates and inlets of his life!

the idea of sickness besieging the body had occurred to him at the time of his great anxiety for Davy.

The friendship between poet and scientist had been constant during Davy's early London years. Their letters provide a commentary on both their lives; Coleridge's notebooks show that he did not attend Davy's first lectures merely to increase, as he said, his stock of metaphors.

There is a strong contrast between the occasional high spirits of Coleridge's earlier letters and the frequent dejection of his later ones, when pain had weakened him and what, in his self-abasement, he called idiocy of the will had begun to affect his whole being. In the earlier letters Coleridge sometimes mentioned his small son, Hartley, for he knew that Davy was fond of children and liked to hear of them. It was in a letter to Davy that Coleridge repeated Hartley's remark about the moon. He was describing the sky from his window at Greta Hall and then recalled a remark of Hartley an evening or two before:

> There is a deep blue cloud over the heavens; the lake and vale, and the mountains, are all in darkness; only the *summits* of all the mountains in long ridges, covered with snow, are bright to a dazzling excess. A glorious scene! Hartley was in my arms the other evening looking at the sky; he saw the moon glide into a large cloud. Shortly after, at another part of the cloud, several stars sailed in. Says he, 'Pretty creatures! They are going to see after their mother the moon.'

On another occasion he set down, in a postscript to one of his letters to Davy, the image of Hartley as a spirit that dances on an aspen leaf.

Because of the pain he suffered, Coleridge, before meeting Davy, had experimented with the remedy commonly prescribed at the time – laudanum – in which opium was the main ingredient. He knew the virtue of it, 'very sovereign to mitigate any pain'. He wrote to his brother of the repose it gave, and of how divine that repose was. He was to express later, as he became enslaved, both the pleasure and the 'life-stifling fear, soul-stifling shame'. Davy had known Coleridge before his genial spirits failed; but he was a friend of his, too, at the beginning of the long years of dejection, when Coleridge is able to describe to him, in words only too telling, the extremes of cheerfulness and dejection from which human beings suffer; 'but so I suppose it is with all of us – one while cheerful, stirring, feeling in resistance nothing but a joy and a stimulus; another while drowsy, self-distrusting, prone to rest, loathing our own self-promises, withering our own hopes – our hopes, the vitality and cohesion of our being'.

Coleridge and Davy wrote to their common friends Poole and Southey their hopes and fears for each other and the dangers each saw in the other's situation in life. Coleridge saw Davy exposed to flattery and to the dissipation of his energies in the immense variety of his interests and his ever-widening acquaintance with all kinds of people. Davy wrote of Coleridge's weakness of will, less than ever, he said, commensurate with his ability. But nearly always they wrote with love and admiration and understanding of each other, though Davy had a less sensitive heart than Coleridge. He did not accurately judge the humility which would make S. T. C.'s own epitaph the most touching in English. Davy's outward-flowing nature, his fierce control of his own faculties when in the stress of discovery, made him impatient of Coleridge's weakness, yet he, Coleridge's younger contemporary, recognized in the poet-philosopher a creating being, and gave him the admiration he needed. We can now read almost everything Coleridge ever wrote; Davy had read only what had been published by 1803 – all the best, admittedly – when he wrote to Poole:

Coleridge has left London for Keswick. During his stay in town I saw him seldomer than usual; when I did see him, it was generally in the midst of large companies, where he is the image of power and activity. His eloquence is unimpaired; perhaps it is softer and stronger. His will is less than ever commensurate with his ability. Brilliant images of greatness float upon his mind, like images of the morning clouds on the waters. Their forms are changed by the motions of the waves, they are agitated by every breeze, and modified by every sunbeam. He talked in the course of an hour of beginning three works; he recited the poem of Christabel unfinished, and as I had before heard it. What talent does he not waste in forming visions, sublime, but unconnected with the real world! I have looked to his efforts, as to the efforts of a creating being; but as yet he has not laid the foundations for the new world of intellectual forms.

Two letters, one written by Davy to Coleridge immediately before Coleridge set out for Malta, and Coleridge's reply to it dated March 25, 1804, best reveal the relation of the scientist to the poet. Davy's letter was written from the Royal Institution:

Royal Institution, Twelve o'clock,
Monday.

My dear Coleridge – My mind is disturbed, and my body harassed by many labours; yet I cannot suffer you to depart, without endeavouring to express to you some of the unbroken and higher feelings of my spirit, which have you at once for their cause and object.

Years have passed since we first met; and your presence, and recollections in regard to you, have afforded me continuous sources of enjoyment. Some of the better feelings of my nature have been elevated by your converse; and thoughts which you have nursed, have been to me an eternal source of consolation.

In whatever part of the world you are, you will often live with me, not as a fleeting idea, but as a recollection possessed of creative energy, – as an imagination winged with fire, inspiring and rejoicing.

You must not live much longer without giving to all men the proof of your power, which those who know you feel in admiration. Perhaps at a distance from the applauding and censoring murmurs of the world, you will be best able to execute those great works which are justly expected from you: you are to be the historian of the philosophy of feeling. Do not in any way dissipate your noble nature! Do not give up your birthright!

May you soon recover perfect health – the health of strength and happiness! May you soon return to us, confirmed in all the powers essential to the exertion of genius. You were born for your country, and your native land must be the scene of your activity. I shall expect the time when your spirit, bursting through the clouds of ill-health, will appear to all men, not as an uncertain and brilliant flame, but as a fair and permanent light, fixed, though constantly in motion – as a sun which gives its fire, not only to its attendant planets, but which sends beams from all its parts into all worlds.

May blessings attend you, my dear friend! Do not forget me: we live for different ends, and with different habits and pursuits; but our feelings with regard to each other have, I believe, never altered. They must continue; they can have no natural death; and, I trust, they can never be destroyed by fortune, chance or accident.

H. DAVY.

Coleridge's reply was immediate:

Sunday, March 25, 1804.

My dear Davy – I returned from Mr Northcote's, having been diseased by the change of weather too grievously to permit me to continue sitting: for in those moods of body brisk motion alone can prevent me from falling into distempered sleep. I came in meditating a letter to you, or rather the writing of the letter, which I had meditated yesterday, even while you were yet sitting with us. But it would be the merest confusion of my mind to force it into activity at present. Yours of this morning must have sunken down first, and have found its abiding resting-place. O, my dear friend! blessed are the moments, and if not moments of *humility*, yet as distant from whatever is opposite to humility, as humility itself, when I am able to hope of myself as you have dared to hope of and for me. Alas! they are neither many, nor of quick recurrence. There *is* something, an essential something wanting in me. I feel it, I *know* it – though what it is I can but guess. I have read somewhere that in the tropical climates there are annuals as lofty and of as ample a girth as forest trees: – So by a very dim likeness I seem to myself to distinguish Power from Strength – and to have only the power. But of this I will speak again: for if it be no reality, if it be a disease of my mind, it is yet deep-rooted and of long standing, and requires help from one who loves me in the light of knowledge. I have written these lines with a compelled understanding, my feelings otherwhere at work – and I fear, unwell as I am, to indulge any deep emotion, however ennobled or endeared. Dear Davy! I have always loved, always honoured, always had faith in you, in every part of my being that lies below the surface; and whatever changes may have now and then *rippled* even upon the surface, have been only jealousies concerning you in behalf of all men, and fears from exceeding great hope. I cannot be prevented from uttering and manifesting the strongest convictions and best feelings of my nature by the accident, that they of whom I think so highly, esteem me in return, and entertain reciprocal hopes. No! I would to God, I thought of myself even as you think of me, but . . .

So far I had written, my dear Davy, yesterday afternoon, with all my faculties beclouded, writing mostly about myself, – but, Heaven knows thinking mostly about you. I am too sad, too much dejected to write what I could wish. Of course I shall see you this evening here at a quarter after nine. When I mentioned it to Sir George, 'Too late', said he; 'no, if it were twelve o'clock, it would be better than his not coming.' They are really kind and good. Sir George and Lady Beaumont. Sir George is a remarkably *sensible* man which I mention because it *is* somewhat REMARKABLE in a painter of genius, who is at the same time a man of rank and an exceedingly amusing companion.

I am still but very indifferent – but that is so old a story, that it affects me but little. To see *you* look so very unwell on Saturday, was a new thing to me, and I want a word something short of affright, and a little beyond anxiety, to express the feeling that haunted me in consequence.

I trust that I shall have time, and the gifted Spirit, to write to you from Portsmouth, a part at least of what is in and upon me in my more genial moments.

But always I am and shall be, my dear Davy, with hope, and esteem, and affection, the aggregate of many Davys,

Your sincere friend,

S. T. COLERIDGE.

Before leaving England for Malta Coleridge showed Lady Beaumont Davy's letter that she might realize, he said, the depth and seriousness of his mind.

The Mercurial Chemist

Davy heard from Coleridge in Malta, and wrote of the letter to Tom Poole:

> I have received a letter from Coleridge within the last three weeks. He writes from Malta in good spirits, and, as usual, from the depth of his being. God bless him! He was intended for a great man. I hope and trust he will, at some period, appear as such.

When Coleridge returned from Malta Davy, whose belief in 'sacred independence' was almost fanatical, did his utmost to persuade Coleridge to give a series of lectures at the Royal Institution. He felt that a hundred guineas earned would do Coleridge good; and that Coleridge's eloquence would do the world good, could the world be induced to come under his spell. Tom Poole snorted. He was fierce against what an Edinburgh reviewer had called the thickly fashionable atmosphere of the Royal Institution. He wrote to Josiah Wedgwood that he should try to dissuade Coleridge from giving the lectures, 'because it will detain him from what is of greater immediate importance; because he will never be ready, and therefore always on the fret; because I think his prospects such that it is not prudent to give lectures to ladies and gentlemen in Albemarle Street – Sydney Smith is good enough for them.' Sydney Smith was making huge audiences rock with laughter; it might have done Tom Poole good, and Davy too, to laugh more. But Davy never liked Sydney Smith, and perhaps wanted Coleridge to off-set him. He wrote urgently to Coleridge, and to Poole. The letter to Poole is of interest both in relation to Coleridge and because it gives Davy's views about his country and his times. After a few particulars relating to himself, he wrote in August 1807:

> If Coleridge is still with you, will you be kind enough to say to him, that I wrote nearly a week ago two letters about lectures and not knowing where he was, I addressed them to him at two different places. I wish very much he would seriously determine on this point. The managers of the Royal Institution are very anxious to engage him; and I think he might be of material service to the public, and of benefit to his own mind, to say nothing of the benefit his purse might receive. In the present condition of society, his opinions in matters of taste, literature, and metaphysics, must have a healthy influence; and, unless he soon becomes an actual member of the

living world, he must expect to be brought to judgment 'for hiding his light'.

The times seem to me to be less dangerous to the immediate state of the country than they were four years ago. The extension of the French empire has weakened the disposable forces of France. Bonaparte seems to have abandoned the idea of invasion, and if our government is active, we have little to dread from a maritime war, at least for some time. Sooner or later our colonial empire must fall, in due time, when it has answered its ends.

The wealth of our island may be diminished, but the strength of mind of the people cannot easily pass away; and our literature, our science, and our arts, and the dignity of our nature, depend little upon our external relations. When we had fewer colonies than Genoa, we had Bacons and Shakespeares. The wealth and prosperity of the country are only the *comeliness* of the body – the fulness of the flesh and fat; – but the spirit is independent of them – it requires only muscle, bone and nerve, for the true exercise of its functions. We cannot lose our liberty, because we cannot cease to *think*; and ten millions of people are not easily annihilated.

Poole proved right over Coleridge and the lectures. He was never ready. At first they were put off because of Davy's illness in the winter of 1807. When he delivered the first in January 1808 it seemed that he might be successful. Poole was half converted to the notion of Coleridge as lecturer. But Davy was weak after his dangerous illness. He was not there to help Coleridge to deliver his lecture in the way he had helped Dalton. From de Quincey we have a description, no doubt exaggerated, but graphic, of the difficulties of getting Coleridge to Albemarle Street and of his performance when he got there. After the second lecture his heart often failed him, and he sent messengers with excuses. It was not easy for Davy. He had found a mode of life in which his talents had full play; Coleridge had not. He tried to help his friend and the friend, hating to be helped and to fail, said many wounding things, wounding because, with his profound insight into other people's weaknesses – as well as into his own – and with his miraculous gift of words, Coleridge always found the shaft that pierced. He edged his words with truth. When, in the old days, he had said that Davy was determined to mould himself on his Age in order to make the Age mould itself on him there was truth in it, more truth than in his other *mot*, when he christened Davy a theo-mammonist, a worshipper of God and Mammon at once. But Coleridge's words were not only deadly when tipped with truth, they had immense carrying power. Eagerly we read every word he wrote, whereas Davy's words have lost their speed. There is a sense in which it may be said that we know Davy only dimly – reflected in Coleridge.

The difficulties of the lectures did not break the link between Davy and Coleridge, a link strong because of the shared passion of their pursuit of knowledge. Coleridge made of his abstruse research an inward, subjective thing, teasing it out, and absorbing it into the very life of his life; Davy would seem, in his laboratory, to have reached what might be described as an objective bliss. The two men are of fundamental interest as contrasts, as men experiencing in different modes. For Coleridge never became anything of a scientist, though he had said in 1800, 'As soon as I have disembrangled my affairs I shall attack chemistry like a shark.' Perhaps it was partly because his affairs never did become disembrangled that he did not carry out his resolve; but not entirely. Coleridge was concerned with knowledge in so far as, through feeling, it concerned the individual soul; Davy sought knowledge which might be used to transform the life of future generations. Whenever in his introductory lectures Davy moved at all in Coleridge's territory, Coleridge was not at ease. He detested what he called the slack contemporary use of the words philosopher and philosophical. Soon, he jeered, there would be philosophical cobblers.

Davy received enthusiastic letters from him on the subject of his projected periodical *The Friend*. Money came to be the trouble. Coleridge was the worst business man in the world, partly because he fancied his own acumen. Davy hoped to benefit him by having the successive numbers of *The Friend* printed by Savage (printer to the Royal Institution). But in the end he supported Savage when a dispute arose between him and Coleridge. Coleridge wrote to Daniel Stuart on March 28th, 1809;

I would not consent to be duped by Savage unto my Ruin, neither Davy nor Bernard have ever got me a subscriber – nor tho' I wrote to request it, have my Prospectuses been suffered to appear at the Royal Institution –. This is Friendship and Gratitude. – Davy's conduct *wounds* me – I had written a long Poem (the only verses I have made for years) wholly on his Genius and great Services to mankind, about six weeks ago – we have been intimate these nine years or more – and often has he declared in times of yore, how much he owed to my conversation and incitements – and yet Savage is the more interesting man, and I and my poor Children are to be beggared rather than he affronted.

So Coleridge and Davy drifted apart, not so much through misunderstandings, and the waning of affection, as through the divergency of their ways of life. Had Beddoes lived longer he might have preserved the link between them; but Beddoes had died in 1808. He had felt keenly his own comparative failure. He judged himself. To Davy he wrote, 'Like one who has scattered abroad the *avena fatua* of knowledge from which neither branch, nor blossom nor fruit has resulted, I require the consolations of a friend.' This appeal for help in his extreme despondency was written in his last letter to Davy, from which Davy quotes when writing to Coleridge of the death of their common benefactor. Davy wrote:

December 27. 1808

Alas! poor Beddoes is dead! He died on Christmas eve. He wrote to me two letters on two successive days, 22d and 23d. From the first, which was full of affection and new feeling, I anticipated his state. He is gone at the moment when his mind was purified and exalted for noble affections and great works.

My heart is heavy. I would talk to you of your own plans, which I shall endeavour in every way to promote; I would talk to you of my own labours, which have been incessant since I saw you, and not without result; but I am interrupted by very melancholy feelings, which, when you see this, I know you will partake of.

Ever, my dear Coleridge,
Very affectionately yours,
H. DAVY.

During the years of their friendship there had been very real communion between Davy and Coleridge. From Coleridge Davy learnt to understand something of the imagination. 'The imagination is the only creative faculty of our nature', he was to write to his wife. From Davy Coleridge learnt something of what a scientist meant by truth. In a notebook he wrote, 'To say that truth is of no importance except in the signification of sincerity is to confound sense with madness and the

word of God with a dream.' He wrote, too, that other men's worlds
were his chaos from which to create. For many years Davy's was one
of his worlds, not so much from which to create as to provide material
for his quest – 'to reduce all knowledge into harmony'.

Before the breach, Coleridge and Southey had sometimes laughed
together over Davy. Southey coined the word *metapothecary*, and
Coleridge wrote to him, 'Do, Southey, keep to your most excellent
word (for the invention of which you deserve a pension more than Dr
Johnson for his dictionary).' It is very comic to hear Coleridge, who
talked so much, arraigning Davy for too much talking. He wrote to
Southey:

> . . . it is *talking, talking, talking* that is the cause of the poison. I defy
> Davy to *think* half of what he *talks* . . . chemistry tends in its present
> state to turn its priests into sacrifices . . . it prevents a young man
> from falling in love. We all have obscure feelings, which must be
> connected with something or other – the miser with a guinea – Lord
> Nelson with a blue ribbon, Wordsworth's old Molly with her
> washing tub – Wordsworth with the hills Lakes and trees, (all men
> are poets in their way, tho' for the most part their ways are *damned
> bad ones*) . . . That to be in love is simply to confine the feelings
> prospective of animal enjoyment to one woman is a gross mistake –
> it is to associate a large proportion of all our obscure feelings with
> a real form . . . A young poet may do without being in love with
> a woman – it is enough if he loves – but to a young chemist it would
> be salvation to be downright romantically in love – and unfor-
> tunately so far from the poison and the antidote growing together,
> they are like the wheat and the Barbary.

During his first ten years in London Davy was convivial, and courted,
if not in love. His various journeys, too, were always in themselves
delightful holidays, leisurely excursions, in which he diverted his steps
to call on his friends, or to see for himself places of which he had been
told, or geological formations he wanted to examine. He came to have
a special affection for Ireland. He said he never entered it without
feeling his spirits rise, partly from the kindness which he experienced
there, and partly from the original and diverting manner of the people.
He went to Ireland in 1805 and again in 1806 when, as was natural,
because of his friendship with Anna Beddoes, he visited her father's
family and her sister Maria. Maria had met Davy in Bristol and again

in London. In London she said she thought he had improved. But whether she thought him improved in manner or modesty is not very clear. She said he was 'not so much of a cosmogony man'. Davy had also met Maria's father, Richard Lovell Edgeworth, in Bristol not long after R. L. had married his fourth successive wife. In all he was to have twenty or so children, so there was an abundance of material for his educational experiments. Davy did not find his endless talk a bore as Byron did; nor was he as sceptical about his theories, based on Rousseau's writings, as was Coleridge, who said Josiah Wedgwood had informed him that the Edgeworths 'were most miserable as children, and yet the father was always vapourising about their happiness'. Davy listened. He was always ready to enter with interest into anyone's ideas. He called Edgeworthstown the 'moral and intellectual paradise of the author of *Castle Rackrent*'. He greatly admired this novel for its Irishness, told Maria an amusing anecdote which had happened to himself, and which she used in *Ennui*, and was ready to help Edgeworth with ideas on the subject of professional education. Maria, a shrewd judge of character, liked her guest. She found him unpretentious and goodnatured, and called on him without compunction in subsequent letters for help when she thought it would aid her father.

In his various visits to Ireland, Davy's comments range from the legends of the people, and accounts of their poverty, to geology. In all his descriptions, his preference for the wild and romantic in natural scenery is evident, and his equally strong preference for the civil in human life. Incentives to endeavour, the introduction of manufactures, increased knowledge, religious tolerance and, following Rumford, the amelioration of man's condition by the study and application of scientific laws – these were Davy's cures for misery. He indulged in what he thought were rational anticipations of man's happiness, while always retaining his insight into the turbulence of man's passions. He recorded the fighting nature of the Irish, and lamented man's political ineptitude.

He set out his views in a letter to Poole. Poole had written to consult him about a mine, and must have added some political remarks. Davy wrote in answer:

My dear Poole,
    What you have written concerning the indifference of men with regard to the interest of the species in future ages, is perfectly just and philosophical; but the greatest misfortune is, that men do not

attend even to their own interest, and to the interest of their own age in public matters. They think in moments instead of thinking as they ought to do, in years; and they are guided by expediency rather than by reason. The true political maxim is, that the good of the whole community is the good of every individual; but how few statesmen have ever been guided by this principle! In almost all governments the plan has been to sacrifice one part of the community to the other parts; sometimes the people to the aristocracy; at other times the aristocracy to the people; sometimes the colonies to the mother country, and at other times the mother country to the colonies. A generous enlightened policy has never existed in Europe since the days of Alfred; and what has been called 'the balance of power', the support of civilisation, has been produced only by jealousy, envy, bitterness, contest and eternal war, either carried on by pens or cannon, destroying men morally and physically! But if I proceed in vague political declamation, I shall have no room left for the main object of my letter – your mine. I wish it had been in my power to write decidedly on the subject; but your county is a peculiar one. Such indications would be highly favourable in Cornwall; but in a *shell limestone*, of late formation, there have as yet been no instances of great copper mines. I hope, however, that your mine will produce a rich store of *facts*. . . .

Had I been rich, I would venture; but I am just going to embark with all the little money I have been able to save for a scientific expedition to Norway, Lapland and Sweden. In all climes,

I shall be your warm and sincere friend,

H. DAVY.

He did not go to Norway and Sweden till a dozen years and more had passed; and he never reached Lapland. He was too busy. During the winter of 1810 he was seconded by the Managers of the Royal Institution in order to lecture on electrochemistry to the Dublin Society. He also, at the request of the Farming Society of Ireland, repeated the six lectures on the application of chemistry to agriculture which he had delivered to the Board of Agriculture in England. In 1811 he gave two courses of lectures in Ireland, one on the elements of chemical philosophy and the other on geology. Not even in Albemarle Street was he received with greater acclamation than in Dublin. Tickets were at a premium; all Dublin opened their doors to him and made him their guest of guests, so that the warmth of his reception surprised and touched him. The Provost and Fellows of Trinity College conferred on him the honorary degree of LL.D.

In one of his introductury lectures he discoursed on the education of women. He had read Mary Wollstonecraft. He urged that women should be rescued from the frivolous type of education; that natural science might play in their training in accuracy the part of mathematics in the education of men – even Milton's Eve had had a chance in Eden! He closed with an eloquent plea that women should be allowed a share in the grand privilege of human nature – the pursuit of knowledge.

Not actually in the audience listening to this peroration, but certain to have it all reported to her, for she had friends in Dublin, and Davy had first met her in Ireland, was Mrs Apreece with whom he had fallen in love.

10 · *Jane Kerr of Kelso*

Mrs Apreece was a young widow of elegance and fortune; small, dark, and very vivacious; a blue-stocking whose genuine sensibilities and aspirations towards learning are so overlaid with the romantic and factitious that it seems almost impossible to divine her true nature however often she flits in and out of the memoirs of the time. Scott – who nevertheless said he would cry heartily if anything were to ail his little cousin – wrote in his *Journal* a penetrating, if, for so good-natured a man, waspish analysis of her character. He considered her a very curious instance of an active-minded woman forcing her way to a point from which she seemed furthest excluded. For, though clever and even witty, she had no peculiar accomplishment, he said, and certainly no good taste by nature either for science or for letters.

She was the daughter and co-heiress of Charles Kerr of Kelso, who had been secretary to Lord Rodney, and who had made a fortune in Antigua. The Scottish cousinship, which Scott claimed, was through Isabel Kerr of the good but decayed family of Kerr of Bloodielaws. On the mother's side Mrs Apreece was descended from the Tweedie family; her mother had been Jane Tweedie, and she herself was Jane. As Jane Kerr she had married, at the age of nineteen, Shuckburgh Ashby Apreece, eldest son of a Welsh baronet; the marriage was not happy and when her husband died in 1807 she, having had no children, must have felt deliciously free and ready to start again.

Already she had travelled widely. In London she was liked and admired; but it was in Edinburgh that she became the toast of the learned world and queen of a coterie. Vivacious is the epithet most usually applied to her; we hear from several sources of her 'sparkling vivacity'. There is little doubt that she could be very entertaining. And she wanted to go everywhere, know everyone, and see everything. Sensibility was all the fashion, especially sensibility to nature and to the tender passion. She went to the Hebrides to be moved by the scenery, and she burst into tears on hearing Tom Moore sing one of his Irish melodies.

Yet Scott could not be persuaded that her passionate responses were quite genuine. It was with him that she went to the Hebrides and he wondered to observe how amidst seasickness, fatigue, some danger, and a good deal of indifference as to what she saw, she gallantly maintained her determination to see everything. He said it marked her strength of character, and added that she joined to it much tact, and always addressed people on the right side. He continued, 'So she stands high, and deservedly so, for to these active qualities, more French I think than English, and partaking of the Creole vivacity and suppleness of character, she adds, I believe, honourable principles and an excellent heart. As a lion-catcher, I would pit her against the world. She flung her lasso over Byron himself.'

It was after her marriage with Davy that she caught Byron. In Edinburgh she practised on the lesser lions with the success of a Zuleika Dobson; almost one and all they were ready to cry, 'Wilt thou set thy foot o' my neck?'; but more especially John Playfair, the well-known mathematician and natural philosopher. Scott thought he had actually proposed to her, but was a little too old. Sydney Smith, among others, extracted considerable amusement from the predicament of the love-lorn mathematician. Mrs Apreece called on Sydney Smith with letters of introduction while he held the living of Foston-le-Clay. She was passing near the village on her way from Edinburgh to London and wanted to hear the famous wit preach. They became great friends; but her first visit is very amusingly described by Saba, Sydney Smith's daughter, in her *Memoir* of her father. Mrs Apreece travelled in state. Saba Smith, or Lady Holland as she afterwards became, writes:

One Sunday (to show the very primitive state of the villagers) just as he (Sydney Smith) was entering the Church, there was a general rush of the clerk, the sexton, the churchwardens, and principal farmers after him, who with agitated countenances explained, 'Please, your honour, a coach! a coach!' My father with a calmness that filled them with a wonder, said, 'Well, well, my good friends, stand firm; never mind, even though there should be a coach it will do us no harm; Let us see.' And certainly a carriage was seen approaching, such as rarely appeared in these parts; and as it advanced rapidly towards the miserable little hovel that had once been the parsonage house, it was discovered to contain a certain very fashionable lady. The lady turned out to be a Mrs Apreece, on her way from Scotland, bringing letters of introduction to my father whom she was anxious to hear preach; and this was the beginning of an acquaintance which afterwards ripened into intimacy, and several of his most amusing letters are addressed to her, under her more celebrated name of Lady Davy. She and Sir Humphry not infrequently put up at the Rector's Head (as my father used to call his house), and no landlord could rejoice more in 'a run on the road', or more cordially welcome the sight of an old friend.[1]

Jeffrey of the *Edinburgh Review* had warned Sydney Smith that Mrs Apreece was accustomed to much flattery, and that her head had been a little turned. But Sydney took to her at once, divined something genuine under the façade and, although he agreed with Jeffrey that excessive flattery had left her a little bemused, said she seemed a friendly, good-hearted, rational woman, 'as much under the uterine dominion as was graceful and pleasing'. 'He hated', he said, 'a woman who seemed to be hermetically sealed in the lower regions.' He made great game of the devotion of John Playfair who happened on one occasion to be at Foston-le-Clay at the same time as Mrs Apreece. He pretended to his daughter that the lady had succumbed to the charms of the learned man. He said it was wrong at her time of life to be circumvented by Playfair's diagrams; but there was some excuse in the novelty of the attack, as he believed she was the first woman that ever fell victim to algebra, or that was geometrically led from the paths of virtue. A squib written by Scott in 1811 recaptures for us something of the spirit, teasing and gay, which marked Mrs Apreece's conquests among the learned men of Edinburgh:

Have you seen the famed *Bas bleu*, the gentle dame Apreece,
Who at a glance shot through and through the Scots Review,
        And changed its swans to geese?
Playfair forgot his mathematics, astronomy and hydrostatics,
And in her presence often swore, he knew not two and two made
    four.

Playfair, unperturbed, wrote a very cordial letter, after Mrs Apreece had been in Edinburgh nearly a year, recommending her to a friend of

[1] *Memoirs of the Rev. Sydney Smith*, by Lady Holland (Saba), p. 108.

his in London, Miss Mary Berry. 'A most charming and amiable woman' he called her in his letter, lamenting that Mrs Apreece was travelling south, away from him. Both Miss Mary Berry and her sister Agnes – the charming Berrys whose society, when they were young gentlewomen, Horace Walpole perfectly liked, after having endeavoured to find such society for threescore years and ten – found the company of Mrs Apreece as entertaining as Playfair had predicted. They took her to call on Mrs Siddons, were much pleased with her conversation and manner, and were soon writing to Playfair that they heard of her agreeableness from all quarters. She could have achieved her ambition of playing a distinguished part in London society had she never met Davy at all. Scott says her fortune, though easy and handsome, was not large enough for her to make her way by showy entertainments. So she took the *blue* line, and by great tact and management actually established herself as a leader of literary fashion.

At the time when she met Davy she had been as much flattered by men as he by women. He had been considerably pursued by those who admired his bright eyes. But when, through the Edgeworths, he was introduced to Mrs Apreece, it is clear from his letters that he became the pursuer and she the pursued. She seems to have had many doubts, and to have hesitated to accept him; whereas he seems to have had no doubt at all that she was the woman he loved and that he could make her happy. Davy's overtures to Mrs Apreece read like a passage of love-at-arms between a pair of Scott's gentlemanly and gentlewomanly lovers. Critics have said that these are not true to life; they are very true to Davy. He begins by offering to lend his lady his copy of Izaak Walton since she was an initiate in the art of angling at this time. His early notes are in the third person, but as he proceeds his very raptures read like Scott. They throw much light on the tastes of the pair, and on romantic scenery and sentiments in general; but it is she who rallies him about being too romantic, not he her. The early notes are much about fishing. In Ireland she fishes for his sake; in London he goes to the opera for hers. He never had the art of amusing letter-writing, so common in his day, and was little given to airy persiflage. He laboured in compliment. He told her he began to like opera from association, and that the same association would, he thought, make him love a desert; it might even perhaps, in a long time, make him a lover of routs. Although he often praised her talents and powers, he tended to be educative; and this perhaps helped her to think she was in love with him; for it was when eloquently lecturing, and when his expressive face was lit with the pleasure of swiftly linked and spoken ideas, that women wished to follow him. They loved him as they might love a preacher.

There were withdrawals on her part, and earnest pleading, and protestations of exclusive devotion on his. She cared for general society and shone in company; he had an enjoying nature, but his passion was for research. She named his love of science her rival. When separated, the two wrote each other long letters. In one of his, Davy mentions Anna Beddoes. He wrote 'I have had a letter from Mrs Beddoes, and one from George Knox. Both speak of you, both write in a most cheerful tone of themselves. Mrs B. says, "I do admire Mrs Apreece; I think her very pleasing, I feel her abilities, and almost believe, if I knew her I should love her; I suppose then she would love me." '

In August 1811 Davy joined Mrs Apreece in Wales, and sailed with her down the Wye to Monmouth. Then he made a two-day journey through Wales to Holyhead where, in October 1811, while waiting to cross to Dublin, he wrote to thank her for her society. After a metaphysical disquisition on the lines of Anacreon – 'It is hard to love, it is hard not to love, but hardest of all to be absent from the beloved object', he continued:

I cannot describe to you the gratitude I feel for the very happy days and hours I passed in your society. I am convinced with you, that pleasure is always useful to a well-organised mind. You have refined many of my sentiments, given me more correct principles of taste, and raised very much my opinion of the standard of human excellence, by displaying a character which appeared more exalted the more it was studied. I hope you will not think this acknowledgement of benefits obtrusive; it is one that I ought to make for I have sometimes said to you that the pleasure I derived from your conversation interfered with my scientific pursuits. I have gained much and lost little. I have gained sentiments which I trust will continue to exalt and improve my mind, which I am convinced will exist under all circumstances, which in prosperity will be a delight, and which, should adversity be my lot, will be a consolation. All this you will say, is romantic, but who would not follow a romance of pleasure? If this is romantic, it is romantic to pursue one object in science, to attach the feelings strongly; – it is romantic to love the good, to admire the wise, to quit low and mean things and seek for excellence.

He went on to describe the glory and sublimity of clouded Snowdonia and ended his letter, this time, with 'Adieu, my charming friend, my dear grandmama [she was two years younger than he], or by whatever tender or kind name you will permit me to call you and believe me to be, in calms and storms, unalterably your devoted admirer, H. Davy.'

This letter, and another which followed it describing the scenery of Ireland, very naturally, I think, gave Mrs Apreece pause. Did she, who knew more light-tongued companionship, not reflect that such a man as this, as a husband, might become a bore? At any rate she delayed to reply, and Davy wrote her a letter which brings us nearer him than any other of his more lengthy epistles. It was written from the Dublin Society and is possessed by the Royal Institution in the envelope in which Mrs Apreece received it. Davy wrote:

It is now a fortnight since in the regular course of the post you ought to have received a letter from me, a long and I fear tedious letter containing some account of the west of Ireland. For the last week I have watched with anxiety the packets hoping, I may say expecting, to hear from you. To state that I have been disappointed would be saying too little. I have really been unhappy. I am not conscious of having written anything which ought to have offended you. I sometimes imagine that you are ill, sometimes that you intend to neglect me. You perhaps will think me equally unreasonable in my hopes and fears; but you know you promised to write to me and that you told me you expected what I cannot give, entertaining letters; and now more than half the time that I am to devote to Ireland is lapsed; and I should have been contented with three lines if they had told me you were well . . .

I have been living in the midst of the most brilliant society of this town; but I have not enjoyed it, not at all since I have been waiting in vain to hear from you. Even Lady Caher has lost much of her power to waken me from uneasy thoughts; yet I have seen her in a more impressive and amiable character than the world will allow she possesses as a kind attentive and enlightened mother. Her little girl is beautiful, arch, full of life. She has won my admiration in spite of my anti-gallican feelings. She does not speak à single word of English.

Mrs Parkhurst complains of you as a correspondent. I believe she is very amiable and she has seen much, heard much and recollected much. She does not seem happy and appears to live more in her friends than her family and I think a few lines from you would give her pleasure.

I will not go on writing as much about people as I wrote about things in my last letter, for I should make this as long and as tiresome. I know from experience that there is no persecution more disagreeable than that of unsought for excess of kindness, and you are the last person in the world that I should wish to inflict any kind of persecution upon. My affection is at least as disinterested as it is deep and it will never ask you for a single sacrifice not even of time. With the same honesty of feeling I speak of my pleasures and pains. Sympathy in joy I should prefer to sympathy in uneasiness from those I love most except when it is in their power to remove it.

Very sincerely your unchanged unspirited

H. D.

The letter is in place as showing one phase in the relationship of the two, and as an expression of an unusual mood of despondency. Davy was exceptional among his associates in seeking to share with his friends his joy and success rather than his discomfitures and grievances. Mrs Apreece was evidently touched by his letter for she did not continue silent. Happy relations were re-established between them; after his final lecture in Dublin, on December 4th he wrote:

You are now at Loughton. I have the power of dreaming and of picture-making as strong as when I was fifteen. I call up the green woods and the gleams of sunshine darting through them, and the upland meadows where we took our long walk. I seem to hear, as then, the delightful song of the nightingale interrupted by the more delightful sounds of your voice. You perhaps will laugh at this visionary mood, and call it romance; but without such feelings life would be of little worth, and neither our affections nor objects of pursuit would be permanent. It is the continuity and unbroken recollection of pleasurable feelings which constitute the strength and vitality of our being. They are to thought what melody is to music. The mind in a healthy state must always blend its new impulses with old affections. Without this, its tones are like those of the Aeolian harp, broken, wild, and uncertain, fickle as the wind that produced them, beginning without order, ending without effect.

Jane Kerr of Kelso

There has been no packet for three days; there is a report of one being lost. We shall depart as soon as the seas are quiet. To see you is the strongest wish of my heart. My last lecture was received with a feeling and effect that over-powered me. It made me feel a deeper responsibility than I ever felt before; and the enthusiastic attentions paid to me, make me humble rather than vain.

His attitude to her was very much the deferential attitude of an Elizabethan lover to his mistress. He wrote her sonnets and sent her mottoes in Latin and Greek, translating the Greek, but trusting to her own Latinity for 'the silver tongue'. Throughout the courtship she feared his devotion to research, and hinted that he must sacrifice every other ambition to obtain her. He wrote:

I hope, my darling friend, that you bear no uneasiness in your kind and good heart and that you give its true meaning to my unlucky sentence. Indeed I never in the whole course of our social intercourse ever intended to offend you or give you a moment of uneasiness and I do not think I should consider anything *long painful* that I thought would promote your happiness even though it should require from me the greatest of all sacrifices. You know what this is and I trust you will never oblige me to make it.

During the early months of 1812, when Davy was working in Albemarle Street and Mrs Apreece was living with her mother at 23 Grosvenor Street, the porter of the Royal Institution was frequently required to convey notes from his street to the so-nearly-adjoining one. Some of these notes found their way back to the Royal Institution and are folded away there, the folds remaining as Davy pressed them. They range through short missives displaying a little scientific wit, notes making appointments or asking for them, to urgent inquiries when illness is causing anxiety. In the earliest notes Davy merely aspires to be 'of her train'; then he becomes more particular:

... Pray do you go to Miss Churchill's tonight or to Miss Hope's tomorrow night? I wish to know as you are my magnet (though you differ from a magnet in having no repulsive point) and direct my course ...

Some notes are addressed to 23 Grosvenor Street, some to 16 Berkeley Square. Propinquity may have had something to do with his success.

His greatest alarm was when Mrs Apreece, who had irritable nerves and frequently an ailing body, fell dangerously ill. He wrote to her:

> For the first time in my life I have wished to be a woman that I might watch at your bedside; I might wish that I had not given up the early pursuit of medicine for then I might have been admitted as your physician . . .

She must have accepted his suit immediately after her recovery at the beginning of March, for in a note to her dated March 2nd he wrote:

> I have passed a sleepless night from excess of happiness. It seems to me as if I began to live only a few hours ago.

And to his brother John, now studying medicine at Edinburgh University, he addressed the following:

> March, 1812.
>
> My dear John,
>   Many thanks for your last letter. I have been very miserable. The lady whom I love best of any human being, has been very ill. She is now well and I am happy.
>   Mrs Apreece has consented to marry me, and when this event takes place, I shall not envy kings, princes, or potentates.
>   Do not fall in love. It is very dangerous!
>   My case is a fortunate one. I do not believe there exists another being possessed of such high intellectual powers, just views, and refined tastes, as the object of my admiration. I am, my dear brother,
>
> Ever most affectionately yours
>
> H. DAVY.

On April 8th, at a levée held at Carlton House, Davy received from the Prince Regent – it was the first honour to be conferred by the Regent – the honour of Knighthood; and on April 11th he was married to Mrs Apreece by the Bishop of Carlisle. Sydney Smith had begged her not to marry; she would, he said, be lost to her male friends and become a neutral salt instead of an alkali or an acid. Many good-natured squibs were written; the best, I think, in view of the fact that Mrs Apreece had never lacked admirers, and that Sydney Smith had said that whenever she appeared, though there was no garrison within twelve miles, the horizon was immediately clouded with majors, is the following:

> To the famed widow vainly bow
> Church, Army, Bar and Navy;
> Says she, 'I dare not take a vow
> But I will have my Davy.'

John William Ward, afterwards to become the first Earl of Dudley, and a great friend of Lady Davy's, had at first written that she was 'fiercely ugly'. But Sir Joseph Banks, giving Sir George Staunton news of Secretary Davy's intended bride, said she was a rich and handsome widow; he feared, though, that she might 'bring him into Parliament and make a fool of him'.

In later years Lydia White – the Miss Diddle of Byron's *Blues* – merrily described Sir Humphry's courtship of Mrs Apreece to Henry Edward Fox. She told him,

He proposed, she refused; he proposed again, she refused again; he vowed he would marry her or leave her for ever. He went to Ireland; she fell dangerously ill; his agitation was great, he said 'he could live in the world without Mrs Aprice [*sic*], but could not after her death'. He returned to England; she asked him to dinner. He answered, 'You know the conditions.' The invitation was repeated. She was Lady Davy in a few days.

11 · Early Days of Marriage: Michael Faraday

Like Anne Elliot and Captain Wentworth, Davy and Lady Davy had one independent fortune between them; but it was hers. Money is alluring. Was Davy mercenary? Few men have rejected greater avenues to wealth. He dreamed he had made a marriage of true minds. He was immensely more sympathetic towards women's desire for knowledge than most of his friends. Coleridge said, 'The longer I live the more do I loathe in stomach and deprecate in judgment *all* blue-stockingism.' Davy wrote, 'The standard of the consideration and importance of females in a society is, I believe, likewise the standard of civilisation.'

Undoubtedly Davy thought that Lady Davy was so sympathetic towards his highest aims that she would support him in them. He said as much to his mother. He told his mother he hoped she would sympathize in his happiness over his marriage. He believed he should never have married, but for this charming woman whose views and tastes coincided with his own, and who was eminently qualified to promote his best efforts and objects in life. She was equally distinguished he said for talents, virtues, and accomplishments; and the object of admiration in the first society. Poor Mrs Davy! What a paragon of a daughter-in-law must she not have thought Humphry had secured for her. What did his sisters and his aunts think? He sent them his kind love. But he never at any time took Lady Davy to Penzance. John was the only one of the family who came to know her. A fortnight after the marriage, Humphry wrote of her to John, as a most exalted and charming intellectual woman, full of good feelings, refined taste, and having a mind stored with various knowledge. He thought he had every prospect of happiness in his new relation. His wife sent her love to John; she had already given him an introduction to the Mackenzie family, and to others in Edinburgh and would, Humphry said, have a regard for him for his own sake as well as for her husband's as soon as she knew him. 'She is a noble creature', he wrote in another letter to John. He reveals what he hoped his union might achieve in the words: 'My usefulness will be her happiness, and her happiness my glory.'

Perhaps marriage with a woman he thought he could love, and one who was also rich, was an insidious temptation. He meant to labour in the cause of science with a zeal not diminished by increase of happiness, and, with respect to the world, increased independence. A glorious prospect it must have seemed to him – to be able to give up routine lecturing; to become, in the company of a wandering literary scholar, a wandering scientist; a little portable chemical apparatus packed in the carriage with them; all the houses of their fishing and shooting friends open to entertain them; and he himself, released from the vexations attendant on the failure of his plans to ventilate the House of Lords, careless of the quick lampoon and the ready gibe, delivered from the platitudes of Sir Joseph Banks, freed from all chagrins, to follow what Napoleon defined as the main business of life – finding out what he didn't know by what he did.

At first his dream seemed likely to be realized. He was married in April 1812 – in the same week as the fall of Badajos, by which the initiative in Spain passed to Wellington; and while Napoleon, failing at last to guess 'what was on the other side of the hill', was taking the route which led to Moscow. He was married in April, and at the beginning of June he was ready to publish his book, *Elements of Chemical Philosophy*. The Advertisement, or Foreword, to the first edition was dated June 1, 1812, from Berkeley Square, dedicated to his wife.

Out-of-date science books, it has been said, are histories of error. Knowledge has made terrifying strides since Charles Lamb forebodingly wrote, 'There is a march of science, but who shall beat the drums of its retreat?' Lamb puts scientific treatises (along with Court Calendars, Directories, Pocket Books, Draught-boards bound and lettered at the back) among those things in the shape of books that he could not allow as such; they were *things* in books' clothing.

Quite other was the youthful Shelley. He made sure that a copy of the *Elements of Chemical Philosophy* should reach him, and he made abstracts from the *Elements of Agricultural Chemistry* published in the following year. Mr Edmund Blunden has indicated that Shelley read

The Mercurial Chemist

Davy. To both, Bacon was a hero, and both had read Erasmus Darwin. In his *Historical View of the Progress of Chemistry* with which Davy opened his book — still readable because of the 'characters' of Black, Priestley, Lavoisier, Cavendish, and others — he wrote of how Bacon had said that the foundations of true systems of knowledge were to be discovered, not in the books of the ancients, not in metaphysical theories, not in the fancies of men, but in the visible and tangible external world. Experiment was, he said, as it were the chain that binds down the Proteus of nature, and obliges it to confess its real form and divine origin. In *The Triumph of Life* Shelley wrote of Aristotle's pupil, Alexander, 'whom fame singled out for her thunder-bearing minion', of Aristotle himself [the 'other' of the line below] and of Bacon:

> The other long outlived both woes and wars,
> Throned in the thoughts of men, and still had kept
> The jealous key of truth's eternal doors,
>
> If Bacon's eagle spirit had not leapt
> Like lightning out of darkness — he compelled
> The Proteus shape of Nature as it slept
>
> To wake, and lead him to the caves that held
> The treasure of the secrets of its reign.

Shelley declared he would rather be damned with Plato and Lord Bacon than go to heaven with Paley and Malthus. Davy had more insight into Bacon — he did not forget his contemning of Gilbert. But it was Shelley who made Bacon's vision, which inspired both Davy and himself, the fire of lyric. One has only to read the song of the Earth towards the close of *Prometheus Unbound* to know that Shelley was kindled by the dream of man's possible dominion through knowledge as Wordsworth and Coleridge, though Coleridge praised Davy as a Baconian, never were.

> The lightning is his slave; heaven's utmost deep
> Gives up her stars, and like a flock of sheep
> They pass before his eye, are numbered, and roll on!
> The tempest is his steed, he strides the air;
> And the abyss shouts from her depth laid bare,
> Heaven, hast thou secrets? Man unveils me; I have none.

Early Days of Marriage: Michael Faraday

Davy had worked at his *Elements of Chemical Philosophy* with his usual rapid concentration. Once it was completed, he set out with Lady Davy on a bridal tour which was in the nature of a Progress. William Ward wrote to Ivy [Mrs D. Stewart] on July 6th,

> You are to have all the world in Scotland this summer. Sir Humphry and Lady Davy set out tomorrow; . . . Playfair I find advances into Yorkshire to meet his former love. Poor Dieb, I am sorry for him. The widow's choice has given Chemistry a most undue triumph over the exact sciences.

In Scotland, Davy met with that hospitality, including good sport, in which his heart rejoiced. Whether Lady Davy's danced to the same tune is not so sure. Her love of angling had been, perhaps, platonic, whereas Davy always fished with the absorption of a madman. John Davy met the pair on horseback and accompanied them for part of the way. When they came within sight of the river Awe, Humphry, longing to get at it, changed places with John, jumping into the saddle while John took his brother's seat in the carriage. Shortly afterwards the horse reappeared, galloping wildly back without its rider and, because of the steep rocky road, terrifying Lady Davy and John who pictured Humphry thrown and dead. But when they turned an angle of the road there he was in the distance serenely pursuing his sport. He had been so anxious to start fishing that he hadn't had the patience to tie the horse up properly. At Dunrobin Castle, in Sutherland, the Marquis of Stafford's place, to which they went, Davy fished the river Brora flowing out of the loch of the same name and had excellent sport. He wrote to John George Children that the house was so delightful, the scenery so grand, and the field sports so perfect, that he and Lady Davy would stay another fortnight.

From Dunrobin they went to the Duke of Gordon's, from there to the Duke of Atholl's, then to Lord Mansfield's (Scone House), and then to Edinburgh. No wonder Davy's letters to his fellow-anglers exult in his triumphs with rod and gun. Grouse at every fifty yards, he notes, in the country round Dunrobin; in the course of his adventures he shot his first buck, and one would have thought he must have grown weary of landing salmon.

From Edinburgh, on his return journey, Humphry wrote to his mother in praise of John, 'I doubt not that he will be all we wish.' John was fulfilling aspirations which had once been Humphry's own. He was always generous towards John when counselling his mother. John had asked for an increase in his allowance, and Humphry urges that it may be granted: 'Sixty pounds more or less may be of great importance now, but of none when he is able to provide for himself' – a point of view few elders take. He wrote to his mother, 'If it is not *entirely convenient* to you, I will do it, and lest it should injure John's feelings of independence, it may appear to come from you.' Nowhere else is the delicacy of his imagination in relation to another person so apparent as in all his dealings with his brother. Like all experienced persons, he saw dangers for his junior which he had disregarded when they threatened himself. He had risked his own health without a qualm at Bristol; but when he heard that John had made some experiments on himself with digitalis and other poisons he wrote:

> I hope you will not indulge in trials of this kind. I cannot see any useful results that can arise from them: it is in states of disease, and not of health, that they are to be used, and you may injure your constitution without gaining any result; besides, were I in your place, I should avoid being talked of for anything extraordinary of this kind, as you have already fame of a better kind, and the power of gaining fame of the noblest kind.

He never forgot that his own early notoriety did him some disservice with those for whose good opinion he would have cared most.

Davy and his wife had intended to prolong their tour until the beginning of December; but Lady Davy fell ill, was feverish, had a swollen foot, and was out of spirits. Davy himself was eager to work. He was preparing his Agricultural Lectures for appearance in book form under the title, *Elements of Agricultural Chemistry*, which he was to publish in March 1813. Lady Davy, on her return to London, recovered her health and spirits, but said she was glad for a time of her own fireside. By the spring of the next year the marriage, to various observers, seemed to be going well. Miss Berry writes in her *Journal* in May, 1813, that she supped with Lady Davy to meet the Princess, and notes that among the other guests Byron was present. This social triumph must have been as gratifying to Lady Davy as his own sporting progress had been to her husband. On May 14th she wrote to Sarah Ponsonby, 'Lord Byron is still here, but talking of Greece with the feelings of a poet and the intentions of a wanderer. He is to have a quiet breakfast here for the purpose of introduction to Miss Edgeworth on Monday, and I expect the sense of the one and the imagination of the other with the genius of my own Treasure to afford a high intellectual banquet.' In Henry Crabb Robinson's diary, May 31st, 1813, is the entry: 'Dined with Wordsworth at Mr Carr's. Sir Humphry and Lady Davy there. She and Sir H. seem hardly to have finished their honeymoon. Miss Joanna Baillie said to Wordsworth, "We have witnessed a picturesque happiness!" ' Like Madame de Staël, Jane Davy had a gift for amusing anecdote; and both were tremendous talkers; but Madame de Staël out-talked them all. Byron once said of her in a postscript to Rogers, 'The Staël out-talked Whitbread, overwhelmed his spouse, was ironed by Sheridan, confounded Sir Humphry Davy, and utterly perplexed your slave.'

Davy had shortened his northern tour partly because he was eager to experiment, with John George Children and Warburton, in Children's laboratory at Tonbridge, on the new 'detonating compound' which Ampère had described to him as having been discovered by Dulong about a year previously. Wellington earnestly desired sappers and miners, and miners need explosives. The general had written to Lord Liverpool on April 7th, 1812, after the fall of Badajos, in terms which remain moving: 'The capture of Badajos affords as strong an instance of the gallantry of our troops as has ever been displayed, but I anxiously hope that I shall never again be the instrument of putting them to such a test as that to which they were put last night . . . When I ordered the assault I was certain I should lose my best officers and men. It is a cruel situation for any person to be placed in, and I earnestly recommend to your Lordship to have a corps of sappers and miners formed without loss of time.'

Davy had for a long time been experimenting with azote (the name given by Lavoisier to what is now called nitrogen) and chlorine. Dulong had effected a compound of the two which exploded with the heat of the hand, so Ampère wrote to Davy. Ampère also said that Dulong, in the course of his experiments, had lost a finger and the sight of an eye. When the complete series of *Les Annales de Chimie* and *Le Journal de Physique* arrived in Britain, together with other foreign journals delayed by the war, Davy searched eagerly for information. Ampère's letter contained no account of Dulong's mode of preparation of the substance, nor any details respecting it. Nor could Davy find further hints in the journals. But by November 16th he wrote to John that he had discovered the mode of combining azote and chlorine. He told John the method, but said it must be used with great caution. It was not safe to experiment with a globule larger than a pin's head. In a letter to Sir Joseph Banks, to be communicated to the Royal Society, he emphasized the extreme caution necessary in operating on the substance; for with a quantity scarcely as large as a grain of mustard seed, as he was experimenting with it, in a way he describes in detail, there was a violent flash of light and a sharp report; the tube and glass he was using were broken into small fragments, and he received a severe wound in the transparent cornea of the eye. He concluded his detailed letter to Sir Joseph Banks by pointing out that the mechanical force of this new compound in detonation seemed to be superior to any other known, and the velocity of its action greater.

Of this accident William Ward wrote in December, 1812, 'I have been to see Sir Humphry Davy, Kt., who has hurt one of his eyes. Some say it happened when he was composing a new fulminating oil, and this I presume is the story which the R. Society and the *Institut Imperial* are expected to believe; others that it was occasioned by the blowing up of one of his own powder mills at Tunbridge; others again that Lady D. scratched it in a moment of jealousy – and this account is chiefly credited in domestic circles.'

In making his communication to Sir Joseph Banks, Davy, his eye being still inflamed and weak, had had to make use of an amanuensis. One of those who helped him was Michael Faraday, a journeyman bookbinder of Davy's generation, but thirteen years his junior. Roughly he was John Davy's contemporary.

Faraday was the son of Margaret Hastwell, a Yorkshire farmer's daughter, said to have been of Irish descent, by her marriage with James Faraday, a neighbouring blacksmith. The Faradays came to Newington Butts, near London, where Michael, the third of four children, was born on September 22nd, 1791. His early childhood was passed in poverty, but not I think in unhappiness. He had time to play. He knew family affection, and was brought up in the further security of a small Christian sect. He belonged to two little worlds. Because his parents were Sandemanians he, too, was a Sandemanian. Godwin said Sandeman was a celebrated north-country apostle who, after Calvin had damned ninety-nine in a hundred of mankind, contrived a scheme for damning ninety-nine in a hundred of the followers of Calvin.

Early Days of Marriage: Michael Faraday

Faraday attended with his family, Sunday after Sunday from baby-hood, the little plain chapel at the end of Paul's Alley in Red Cross Street. It must have been something like the chapel described in *Silas Marner*, the chapel in Lantern Yard, except that the Sandemanians had no minister. Expulsion from the fellowship, or even separation from it as necessitated sometimes by travel, gave the anguished sense of loneliness, of being cut off, described with great feeling by George Eliot.

Michael Faraday's birth and upbringing were as broadly different from Davy's as the education and birth of both were different from, say, Henry Cavendish's. No class has a monopoly of genius; and it seems that no form of education is much more sure of fostering it than another. Faraday's childhood, though he suffered privation because of the illness of his father, was happier than Cavendish's, though not ecstatic like Davy's. Margaret Faraday, like Grace Davy, must have been a woman not only of deep piety but of stoic endurance; her life was physically hard in a way unknown to Grace Davy.

Having learnt what he could at a school for the poorest, Faraday at thirteen became general factotum and newspaper boy to Mr Geo. Riebau to whom, a year later, he was apprenticed as a bookbinder – a good trade. His hours of work, though long, must have been not unleisurely, for he found time to read some of the books he bound, among them Mrs Jane Marcet's *Conversations in Chemistry intended more especially for the Female Sex*, on which, he said, he founded his first knowledge of chemistry. Faraday longed for more time to read. To his friend Benjamin Abbott, with whom he began and long maintained an improving correspondence, he wrote: 'Time, Sir, is all I require, and for time will I cry out most heartily. Oh that I could purchase at a cheap rate some of our modern gents' spare hours, nay days; I think it would be a good bargain both for them and me.'

Perhaps it was some of M. Masquerier's time he longed to annex. Refugees tend to have time on their hands, and M. Masquerier, an artist who had escaped from France to England, must have seemed a king of infinite hours to Faraday. Masquerier lodged with Riebau over the bookseller's shop. He taught Faraday something of drawing. In 1809 the Faraday family moved to Weymouth Street, near Portland Place, where, after the death of James Faraday, Margaret Faraday, like Grace Davy when she became a widow, took lodgers.

[135]

Her eldest son, Robert, followed his father's trade of blacksmith, and it was he who paid a shilling a time – a shilling in proportion to his wages must have meant sacrifice – for his younger brother to attend Mr Tatum's lectures on Natural Philosophy. These were private lectures given by Mr Tatum at eight o'clock in the evenings at his house, 52 Dorset Street, Fleet Street.

Towards the end of his seven years' apprenticeship Faraday began to grudge the time given to his trade and to suspect the continued effect of it on his moral nature. He was so uncompromisingly moral that one longs for the devil to be a little more ingratiating with him. He wrote to Sir Joseph Banks asking for some advice, some help, some indication of how to begin to work for science, and must have awaited an answer with such hopes and fears as Hardy has made vivid for us in creating Jude nearly a century later. Faraday's first effort met with the same fate as Jude's. There was no reply from Sir Joseph Banks.

Faraday's apprenticeship ended. On October 12, 1812, he left Riebau to become journeyman to a French refugee named De la Roche, a very different person from Riebau. But a customer of Riebau's, a member of the Royal Institution, whose name was Dance, had given Faraday tickets for Davy's last set of lectures. Faraday made the fullest possible notes of the lectures with drawings of the experiments; the beautifully neat though sometimes ill-spelt manuscript of 386 small quarto pages, still to be seen at the Royal Institution, he then bound and, encouraged by Dance, sent the book to Davy together with a letter stating his longing to escape from trade and enter the service of science in however humble a capacity. In writing the letter the deferential mode of the time and the circumlocutions he was patiently acquiring were at odds with his native directness. But he wrote the kind of letter he felt it was necessary to write. In reply Davy wrote to him, on Christmas Eve, 1812, a letter which Faraday was to treasure to the end of his days. It read:

December 24th, 1812.

Sir,

I am far from displeased with the proof you have given me of your confidence, and which displays great zeal, power of memory, and attention. I am obliged to go out of town, and shall not be settled in town till the end of January: I will then see you at any time you wish.

Early Days of Marriage: Michael Faraday

It would gratify me to be of service to you. I wish it may be in my power.

I am, Sir,

Your obedient humble servant,

H. DAVY.

Because of the weakness of his eye after the accident in the Tonbridge Laboratory, Davy first employed Faraday occasionally as his assistant in writing, advising him not to give up bookbinding, giving him some binding for the Royal Institution, and recommending him to his friends. He said Science was a hard mistress and smiled at the idea that it would be so much better for Faraday's moral nature than bookbinding. Then Davy had occasion suddenly to dismiss William Payne, whose shortcomings are made clear in a note in Davy's hand written at an earlier date in the *Journal* of the Institution (Plate 4, p. 81).

The note was written while Edmund Davy was Davy's assistant and W. Payne the factotum. It read:

Objects much wanted in the laboratory of the Royal Institution: cleanliness, neatness and regularity.

The laboratory must be cleaned every morning when operations are going on before ten o'clock.

It is the business of W. Payne to do this and it is the duty of Mr E. Davy to see that it is done and to take care of and keep in order the apparatus.

There must be in the laboratory pen, ink, paper and wafers, and these must not be kept in the slovenly manner in which they are usually kept. I am now writing with a pen and ink such as was never used in any other place.

The laboratory is constantly in a state of dirt and confusion.

There must be a roller with a coarse towel for washing the hands and a basin of water and soap, and every week at least the whole morning must be devoted to the inspection and ordering of the voltaic battery.

The notes reveal Davy. One hardly hears his natural voice more plainly than in 'I am now writing with a pen and ink such as was never used in any other place.' When W. Payne was dismissed it was of Faraday that Davy thought. Faraday was invited to call at the Royal Institution, was interviewed by Davy, was offered William Payne's job and accepted it. The entry in the Minutes of the Meeting of the Managers of the Royal Institution on March 1st, 1813, reads: 'Sir Humphry Davy has the honour to inform the Managers that he has found a person who is desirous to occupy the situation lately filled by William Payne. His name is Michael Faraday. He is a youth of twenty-two years of age. His habits seem good, his disposition active and cheerful, and his manner intelligent. He is willing to engage himself on the same terms as those given to Mr Payne at the time of quitting the institution.' It was resolved by the Managers that Michael Faraday should be engaged to fill the situation lately occupied by Mr Payne, on the same terms. The terms were 25*s*. a week and the occupation of two rooms at the top of the house in Albemarle Street – a pleasant lodging. Faraday gave proof of the genuineness of his intense desire to follow science. He had just concluded a long apprenticeship to a craft attractive in itself and always in demand.

The education of his hands through bookbinding must have been of service to Faraday in the delicate and skilful manipulation of instruments. He was to be a meticulously neat man, ordering all things as exactly as the pages of the books he bound. He systematized; but he was to be much more than a systematizer. Just as Davy unexpectedly combined with electric qualities of mind great tenacity of purpose, so Faraday with order and perseverance combined immense power of intellectual energy. I suppose it might be said that what Davy once imagined, Faraday proved; but it almost seems as though the combination of Davy and Faraday is an example of two and two making five. Davy once wrote in a notebook, 'No man yet ever made great discoveries in science who was not impelled by an abstracted love.' The two men seem to have had this 'abstracted love' in equal measure.

Faraday's notes had been made of Davy's last regular course of lectures at the Royal Institution. But it was not until the April immediately following Faraday's appointment to William Payne's place that Davy in the general Monthly Meeting of the Members formally resigned his professorship. In begging leave to resign his situation of Professor of Chemistry, Davy said he by no means wished to give up his connexion with the Royal Institution, as he should ever be happy to communicate to it his researches, and to do all in his power to promote its interest and success. When he had retired,

Early Days of Marriage: Michael Faraday

Earl Spencer, in the Chair, moved 'That the thanks of this Meeting be returned to Sir H. Davy for the inestimable services rendered by him to the Royal Institution.' He further moved, 'That in order more strongly to mark the high sense entertained by this Meeting of the merits of Sir H. Davy, he be elected Honorary Professor of Chemistry' which, on being seconded by the Earl of Darnley, met with unanimous approbation. William Brande was elected to the vacant Professorship of Chemistry on June 7th, 1813.

It must have been an anxious time for Faraday. After his removal to the Royal Institution he continued to write long letters to Benjamin Abbot, determined by this exercise to improve his manner of expression. He listened to music – unlike Davy he had a good ear; already he had listened to a good many lectures and watched audiences – the polite, the vulgar, the gazers, the listeners, and the bees of business. He despised, in lectures, 'ale-house tricks', and claims for indulgence when no indulgence was needed. He writes at great length like a wise owl; but the joy of Faraday is that he himself perceived when he was being like a wise owl. We can forgive his heavy step in prose – and few have begun with a heavier – when we find him remarking of himself: 'As when on some secluded branch in a forest far and wide sits perched an owl, who full of self-conceit and self-created wisdom, explains, comments, condemns, ordains and orders things not understood, yet full of his importance still holds forth to the stocks and stones around – so sits and scribbles – Mike.'

He must have worked in the laboratory with Davy during March and the first week in April; for Humphry wrote to John that both he and Jane had greatly enjoyed the month in London, but that also he had been hard at work. He added that his eye was better, and that Jane was well, and that he was off to Penzance, going with a pretty large party into the West and fishing on the way. Blake, Warburton, Pepys, and the Sollys would be with him, combining mineralogy with fishing; he wished John could have been with them. The Cornish journey would be too rapid for Jane, and too interrupted.

He set out in the highest spirits, convinced that the real use of science was in discovery, with all kinds of ideas dazzling his mind; his ten years of compulsory lecturing were over, and their fruits stored in the *Transactions* and his two books. *Elements of Agricultural Chemistry* was just out. He had had a thousand guineas for it, and was to have five hundred for each new edition.

[139]

Science had been raised by him to such an eminence that Byron was writing to Miss Milbanke at about this time that poetry was not high in the scale of intellect: 'I prefer the talents of action – of war, of the senate, or even of science.' The *even* is there, but so is science. Davy has often been held up as a warning, as an example of one whose eyes and mind grew sharp to worldly things; but it was never mere personal aggrandizement he craved. The results of science had to be made known; discoveries used. He did not want to be the man to use them, but he wanted to know those who did. He wrote that had any 'philosopher', that is scientific man, with a philosopher's height and width and breadth of thought, had even the most modest scientific practitioner been in a position, for example, to advise those who planned the Walcheren expedition (1809), the poisonous nature of the air would have been revealed and disaster avoided.

From Andover he sent Jane a fish. In Devon he and his companions still fished, though the Devon rivers afforded little sport. We hear of the travellers at Launceston and Lostwithiel, but I have seen no letters telling of their doings at Penzance. Betsy was delicate at this time, and Grace nervous at the possibility of her brother's going into France. For, strange as it may seem to our notions, Napoleon, informed by the *Institut* that Davy considered his discovery of 1807 might have bearings on the understanding of volcanic action, and that an investigation of the subject might place new powers within man's reach, directed the French Government to give the English scientist unconditional permission to visit Paris and travel through France in order to examine the extinct volcanoes of the Auvergne and the active volcanoes about Naples. The permission was never withdrawn. Davy's aim was to travel for a year, or perhaps two, in Europe and the Near East, taking with him his portable chemical apparatus and conversing with his contemporaries in science. Foreign laboratories were to be open to him. Jane, who had already travelled and who was, like so many of her friends, in love with Italy, was to accompany him; and he invited young Michael Faraday, who had only been in the Royal Institution seven months, and those chiefly months of vacation, to go with him as his assistant in writing and in experiments. A complicating factor arose when Davy's valet, whose office was to have been rather that of courier, was so frightened by his wife's fears that at the last moment he refused to leave England. Davy with reluctance, and apologizing for the necessity, asked Faraday to act in his place until a suitable man could be engaged in France. Faraday's pride was hurt, but he consented, and was in an anomalous position from the start.

It was small wonder that the fears of the valet's wife kept him at home and that Faraday, when the time came for him to leave his mother, half wished he were an insulated person so that he could join Sir Humphry on his travels without the pains of leave-taking. For the allies were closing in on France. On the last day of August 1813 Wellington had taken San Sebastian by assault; on September 3rd he had had news that Austria had joined Russia and Prussia against France; on the 7th of October he won the Battle of Bidassoa and was inside the frontier. And on the 13th October, with Wellington about to invade France from the one side and the allied armies preparing for invasion from the other, Davy and his party left London for Plymouth, the first stage of their journey to Paris.

12 · *The First Continental Journey*

It was a strange fate that brought into close association Jane Davy and Michael Faraday. Neat in dress, in person plain, brought up in a narrow sect, prolix and earnest, incapable of flattery, saved from pedantry only by the humorous eye he could cast upon himself, Faraday was a person the least likely to display qualities pleasing to a woman of fashion. In the passport which he obtained in Paris he described himself as having a round chin, a brown beard, a large mouth, and a great nose; he always had to bespeak his hats because of the size of his head. It was hardly likely that Jane, who had wit without humour, would penetrate to that 'Mike' in Faraday who had humour without wit. Since he was entirely innocent of the ways of travellers and knew no French, he must at first have been a liability and cause of anxiety to Davy rather than an asset. His insularity must have been at least as maddening to Jane Davy's irritable nature as her assumption of superiority was galling to his self-esteem. Jane took her maid, but there is no news of her in the journals or in letters.

Both Davy and Faraday kept journals, and Jane accumulated anecdotes to last her for years. Faraday is most vivid in detailing the modes of conveyance, for he was so unused to travel that things which seemed ordinary to accustomed travellers were extraordinary and interesting to him and, through the lapse of time and change of vehicles, to us. On the day of setting out, Wednesday, October 13th, he wrote, 'Never before, within my recollection, left London at a greater distance than twelve miles. 'Tis indeed a strange venture at this time, to trust ourselves to a foreign and hostile country, where also so little regard is had to protestations and honour that the slightest suspicion would be sufficient to separate us forever from England.' The party travelled in their own carriage, and in two days had reached Plymouth, Faraday recording in his journal, 'A revolution in my ideas respecting the earth's surface.' He was especially taken with the 'mountainous' character of Devon.

[142]

The First Continental Journey

At Plymouth the party embarked in a cartel for Morlaix in Brittany. The carriage had to be taken to pieces for transport and, when the vessel neared Morlaix, Faraday, who had thought to land at once, was filled with indignation to find that they had to wait for a barge full of officers, beggars and porters to come aboard, and were then searched from their hats to their shoes. With British phlegm and heavy sarcasm, Faraday watched the hurry and bustle as the various parts of the carriage, together with the boxes and packages, were brought on deck and lowered into the barge. It was all done so ineffectually to Faraday's eye that sometimes nine or ten men would be round a thing of a hundred pounds weight, each most importantly employed; yet the thing would remain immovable until the crew were urged by their officers or pushed by the cabin-boy. From the seat-boxes of the carriage Davy's party were allowed to take what was absolutely necessary for the night, and next morning they passed the customs. The customs officers ranged themselves on the edge of the quay, and then some thirty or forty inhabitants of the town ran and tumbled down the steps, leapt into the barge, seized some one thing and some another, and conveyed them to the landing-place above. To Faraday it seemed as if a parcel of thieves were scampering away with what was not theirs. He saw no cranes or substitutes for cranes on the quay; the body of the carriage had to be raised to the landing-stage from the barge by hand labour; no order, no method, no regularity, as far as he could see, and yet the work was done and Davy's fears of resigning his carriage to its fate at the foot of the steps were unfounded.

The carriage was searched in every nook and cranny, and thumped in every part for hollow and secret places. In the customs house, where the trunks were operated on, package after package was opened, roll after roll unfolded, each pair of stockings unwrapped, and each article of apparel shaken. Two or three dozen pairs of Lady Davy's white cotton stockings narrowly escaped seizure because they were new, and it was a long time before the arguments of their being necessary for a long journey and of their being marked secured their surrender again. There is a world of meaning in Faraday's use of italics for the word *polite* when he writes that, all being ended, Davy made the customs officers a gift for their *polite* attention.

And then the putting together of the carriage and the replacing of the goods! Faraday was astonished at how, with their poor means, and their want of acquaintance with such affairs, these poor Frenchmen were able to get it into order. True, they made the job appear a mighty one, but they got through it. After exclaiming '*levez! levez!*' for an hour or two, the excitable creatures so managed that everything was in a movable state. Horses were 'tied to', and the party proceeded to the best hotel in the place, where Faraday was aghast at the one entrance to the *cour*, paved similarly to the street, through which horses, pigs, poultry and human beings passed indiscriminately; and at the kitchen fire where beggars and nondescripts of the town warmed themselves and chattered to the mistress while the cooking was in process. Their dishes, Faraday admitted, were excellent and inviting to the taste; but whilst at table it was necessary to dismiss all thoughts respecting cookery and kitchen.

On October 23rd the party set out for Paris. Faraday gives the French postilions a paragraph to themselves. They were mostly young and lively men. Dress varied, but hairy jackets, often finely ornamented, were frequent, and many post-houses maintained a kind of uniform of one colour turned up at the edge with another. Faraday saw his first pair of jack-boots come out of the kitchen at Morlaix. The postilion had left them in the hotel till all was arranged in the carriage, then he used his reserve strength and showed them off in a walk from the fireside to the carriage. Sometimes these boots, weighing from eighteen to twenty pounds, were attached to the saddle by straps, and then the postilion jumped on to his horse and into them at the same time. Their use, according to the wearers, was to save the legs from being broken should the horses stumble or the carriage be overturned. On Sunday, the 24th, at about seven o'clock, the night being very dark, one of the horses of Davy's carriage tumbled over, and while the postilion was mending the traces Faraday saw for the first time a glow-worm, and in a letter home recalled the wonder of it.

He was also amazed at the French pigs at Drieux, pigs so different from the fat English sows that he hardly recognized them as pigs at all. What in the distance he judged a greyhound he was obliged, on a near approach, to acknowledge a pig with a long lean body, back and belly arched upward, lank sides, long slender feet, and capable of outrunning the horses for a mile or two together. The postilion's whip was a most tremendous weapon to dogs, pigs, and little children. With a handle of about thirty inches long, it had a thong of six to eight feet in length, and was constantly in a state of violent vibratory motion over the heads of the horses, giving rise to a rapid succession of stunning sounds.

The First Continental Journey

In Paris, where the Davys put up at the Hôtel des Princes, Rue Richelieu, the travellers met with difficulties and mortifications as well as pleasures. It was not permitted to any but the inhabitants of Paris, whose names were registered as such, to be in the city without passports. When Faraday went to the Préfecture de Police to obtain one, he found the place on the bank of the river, an enormous building, with an infinity of offices. There were a great number of people, but Faraday found his was a peculiar case. Excepting Davy's there was not another free Englishman's passport down on the books. An American who was there and who, perceiving Faraday at a loss for words, spoke to him, could scarcely believe his senses when he saw an official make out a paper for a free Englishman. He would willingly have been inquisitive, but Faraday was equal to him and went off with his passport. Faraday was close by nature, and everything in Paris made him suspicious. He wrote that he was out of patience with the infamous exorbitance of these Parisians. He learnt little of the language. The churches offended him: it could hardly be expected, he said, that they would have attractions for a tasteless heretic. Perhaps Jane Davy had called him a tasteless heretic, for the remark reads like a quotation. He found Mass to be like a play: 'a theatrical air spread through the whole, and I found it impossible to attach serious or important feeling to what was going on'. The spoils of Napoleon's conquests were everywhere displayed; Faraday saw the Emperor himself, sitting in one corner of his carriage, covered and almost hidden from sight by an enormous robe of ermine, and his face overshadowed by a tremendous plume of feathers that descended from a velvet hat. His carriage was very rich, and fourteen servants stood upon it. In Notre-Dame, Faraday and the Davys saw displayed, with the crown and imperial regalia, Napoleon's coronation mantle which was said to weigh more than eighty pounds, and to be lined with the skins of six thousand ermines.

Less insular than his young laboratory assistant, and moving in a different sphere, Davy was hardly less prickly in Paris. He might talk to Poole about the republic of men of science, but he was an intensely patriotic Englishman. The long letter written by him after the final fall of Napoleon, and probably intended for Lord Liverpool, explains his attitude in Paris, and particularly his attitude towards those treasures in the Louvre which had been looted from Italy and Greece. He strongly argued that these possessions, more important, he considered, than territorial ones, should be restored to enrich their native countries and not allowed to remain as the reward of, and enticement to, aggression. Paris, he thought, had no claim to the dignity of being the capital of the arts; he urged that the King of France should command the removal of spoils because 'the first wish of a man, who, after being driven from his house by robbers, finds it crowded on his return by goods plundered from his neighbours, ought to be to make full restitution'. So, in the autumn of 1813, Davy strode through the Louvre muttering, 'What an extraordinary collection of fine frames!' and Faraday, when he went to see for himself, said 'Thieveries!' Davy was accompanied on his tour of inspection by Thomas Richard Underwood, his old acquaintance of Bristol and early London days, whose prudent love query had moved Coleridge to such immoderate mirth. Davy had called on Underwood soon after his arrival in Paris. Underwood, always something of a Jacobin and now a French partisan, had lived in France since the time of the Treaty of Amiens, in theory a prisoner on parole but, because of the favour of the Empress Josephine, who protected him, and to whom he continued to pay court after the divorce, enjoying in practice the freedom of Paris. In the old days at Kynance it was Underwood who had sneered at Davy's raptures. Now the situation was reversed; now Underwood was the rapturous cicerone, and Davy the unmoved initiate, though once he exclaimed, 'Gracious powers, what a beautiful stalactite!'

Often politically irate and socially astonished, Davy was entirely open to delight when it came to fruitful interchange of ideas with the French scientists. Immediately after his arrival in Paris, Ampère, whom the English chemist had chiefly longed to meet, came especially to Paris to see him. Writing to his brother John, Davy said he received every attention from the scientific men of Paris and most liberal permission from the Government to go where he pleased. He lived, he said, very much with Berthollet, Cuvier, Chaptal, Vauquelin, Humboldt, Guyton de Morveau, Clement, Chevreul, and Gay-Lussac, of whom he was to write later, 'Gay-Lussac was quick, lively, ingenious and profound, with great activity of mind, and great facility of manipulation. I should place him at the head of the living chemists of France.'

Davy was always generous in his appreciation of Gay-Lussac's powers, though he came into rivalry with him in relation to the substance which Gay-Lussac had named *iode* from its violet colour in a gaseous state. Davy renamed it *iodine*. Manufacturers tend to be indifferent to questions of priority in discovery; to the hot, injured feelings of rival theorists, and even to questions of national prestige. Their desire is for definite knowledge, however derived, that they may put it to use. Until Davy's arrival, the knowledge of the French chemists concerning the 'new substance', discovered accidentally about the beginning of 1812 by Courtois, a manufacturer of saltpetre, had not been clarified, although Clement, Courtois, and Gay-Lussac had ascertained many of its qualities. Clement asked Ampère to give Davy a small packet of the new substance for examination. Davy, with the use of his portable laboratory at the Hôtel des Princes, satisfied himself in a few days that iodine was in itself a simple substance, analogous in its chemical relations to chlorine. It was because of this analogy that he named it *iodine* in a communication which he subsequently made to the Royal Society. Davy immediately communicated his own results to Gay-Lussac, and seems to have made full acknowledgement of Gay-Lussac's prior work in his own subsequent researches. But both chemists felt aggrieved the one with the other – Gay-Lussac because of the speed with which Davy solved a problem which had occupied the French chemists for months, and Davy because Gay-Lussac published, without acknowledgement, work founded on what he had first learned from Davy. Gay-Lussac did not approve of Ampère's action in calling in Davy at all. Humphry wrote to John Davy, 'I have worked a good deal on iodine and a little on the torpedo. Iodine had been in embryo for two years. I came to Paris; Clement requested me to examine it, and believed that it was a compound, affording muriatic acid. I worked upon it for some time and determined that it was a new body, and that it afforded a peculiar acid by combining with hydrogen, and this I mentioned to Gay-Lussac, Ampère, and other chemists.' Gay-Lussac, Davy added, 'immediately took the word of the Lord out of his servant's mouth' and treated this subject in accordance with the fresh direction given. John Davy says that in less than twelve months, in consequence of the elaborate and masterly researches of Gay-Lussac, the chemical history of iodine was more full and complete than that of most other substances. Davy had given a fresh turn to, and speeded up, the chase at a time when, as far as one can see, all the French dogs were on the wrong scent. It was hard that they should then resent the interference they had invited and made use of. But that was often Davy's fate – to sow a seed from which others gathered the fruit, and it must have been maddening to anyone not a saint. But then he, too, found his own mind fertilized by other men's suggestions. He told John that iodine was as powerful an ally as any he could have found at home. Mind speeds mind – 'it is all a purchase, all is a prize'; jealousies fade, but the fraction of fresh knowledge works, and continues working, in the sense that yeast works.

What Davy writes of Laplace, whom he met in 1813 and again in 1820, is interesting not only in itself but because of its relation to Davy's own estimate of John Dalton; and for the impression it gives of Davy in 1813, not as a man assured and large but as being, in his own idea of himself, 'a young and humble aspirant to chemical glory': Of Laplace he wrote:

I remember the first day I saw him, which was, I believe, in November 1813. On my speaking to him of the atomic theory in Chemistry, and expressing my belief that the science would ultimately be referred to mathematical laws, similar to those which he had so profoundly and successfully established with respect to the mechanical properties of matter, he treated my idea in a tone bordering on contempt, as if angry that any results in chemistry could, even in their future possibilities, be compared with his own labours. When I dined with him in 1820, he discussed the same opinion with acumen and candour, and allowed all the merit of John Dalton. It is true our positions had changed . . . From a young and humble aspirant to chemical glory, I was about to be called, by the voice of my colleagues, to a chair which had been honoured by the last days of Newton.

In 1813 Davy was an enemy national; not in 1820.

If French propaganda annoyed him he marked his annoyance by action. He walked out of the Théâtre de la Porte Saint-Martin in the midst of the performance, when he found that the play was a melodrama in which Lord Cornwallis was represented as the murderer of the children of Tippoo Sahib, and in which the whole action was so directed as to make the English odious in the eyes of the mob. He was not presented to the Emperor. Lady Davy said he objected to attend the levée of his country's bitterest enemy; others said the Emperor expressed no wish to meet the English chemist. Both Davy and his wife were introduced to the Empress Josephine at Malmaison and Davy seems to have been as difficult as possible on this occasion. He made a fuss about clothes, saying, when remonstrated with, 'I shall go in the same dress to Malmaison as that in which I called upon the Prince Regent at Carlton House.' With bad grace he consented to exchange a pair of half-boots that laced in front, and came over the lower part of his pantaloons, for black silk stockings and shoes. Lady Davy's dress, too, caused some difficulty. She went to the levée lightly clad, although the date was November 30th. After being greeted with the others present in the Salle de Réception, Davy and Lady Davy were invited by the Empress into the Boudoir, where Lady Davy was presented with a china cup which she admired. When she said she thought herself not dressed warmly enough to visit the conservatories, the ladies-in-waiting were commanded to bring cloaks. According to Underwood, who was present, a mountain of the most costly and magnificent furs was displayed before the guest, the splendid trophies, he concluded, of the royal reconciliation at Tilsit.

On an earlier occasion Lady Davy and her maid had been mobbed in the Tuileries. The French ladies were wearing such enormous bonnets that Lady Davy's little cockle-shell of a hat, the latest fashion in London, attracted the attention of a holiday crowd who gathered round to gaze at the marvel. One of the custodians of the gardens appeared and told the English lady that no cause of *rassemblement* could be allowed, and that she must withdraw. She appealed to some officers of the Imperial Guard, who regretted that they could do nothing to prevent the order from being enforced; but one of them offered the lady his arm to conduct her to her carriage. By this time, however, the crowd had so much increased that the affair began to look ugly. It became necessary to send for a corporal's guard, and the party left the gardens surrounded by fixed bayonets.

One old friend, in addition to Underwood, whom Davy looked up was Count Rumford, who had been partly instrumental in bringing him into the Royal Institution. Rumford had never returned to London after leaving it in 1802. He had married Lavoisier's wealthy widow, a woman of charm and high intelligence who, according to Mary Sommerville, became, as she grew older, capricious and ill-tempered. Rumford had obtained a divorce in 1809. He understood things better than people, being too much devoted to order to allow for human unpredictability. The work on which he was engaged when he died was *The Nature and the Effects of Order*. With all his good designs for social reform, he tended to think of people as abstractions, an assumption that is death to humanity and art. As Godwin said to Mary Shelley when criticizing her play for the mathematical abstractness of her characters, 'If A crosses B, and C falls upon D, who can weep for that?'

Davy talked to Rumford of Faraday and praised him highly. He had not brought him to Auteuil, fearing Rumford might not care to see a stranger. When Davy spoke of the courtesy of the French scientists Rumford assured him he would not have been so well treated had he been staying in Paris for good. After Rumford's death his only daughter, Sally Thomson, sent Davy her father's gold watch and chain, thus carrying out his expressed wish.

On December 13th Davy was elected a corresponding member of the *Institut Imperial*. He and his party left Paris on December 23rd.

By December 29th they were in Fontainebleau, where Davy walked in the forest and, visiting the palace, had a strong presentiment of the fall of Napoleon – it was at Fontainebleau that the Emperor was, later, forced by his marshals to abdicate. The winter evening, the departing year, the impending fall of the golden eagle, even though he saw it shine as a bird of prey, 'Emblem of rapine and of lawless power', induced in Davy a mood of poetry. In that season of the sleep of things he saw the forest clothed with a magic foliage of ice. The oaks were girt round with massy ice, the birches and beeches glittered with sunset colours; the huge rocks rose as if the work of human art. Davy felt the 'fitful change' of all human things:

> An empire rises, like a cloud in heaven,
> Red in the morning sun, spreading its tints
> Of golden hue along the feverish sky,
> And filling the horizon; – soon its tints
> Are darkened, and it brings the thunder storm, –
> Lightning and hail, and desolation comes;
> But in destroying it dissolves, and falls
> Never to rise!

Faraday, too, struck by the icy glory of the scene, described it in his journal, and with the word 'airiness' conveys something which completes the vision: 'Every visible object', he writes, 'was resplendent in a garment of wonderful airiness and delicacy.'

The party had driven from Fontainebleau to the mountainous region of Auvergne, visiting the most remarkable extinct volcanoes in the south of France. At Montpellier, where they remained a month, Davy worked both in the laboratory of M. Berard and with his own apparatus on the subject of iodine, and sought for traces of it in the seaweeds of the Mediterranean. With M. Berard he went to see Vaucluse and was enraptured by it. His verse description of the scene conveys the impression of a mind 'alive to sympathy with all created forms'; yet that fortunate magic that makes immortal verse still eluded him. Immensely stimulated and excited by the beauties he saw, perhaps he tried too often and too hard. He could never let himself rest. He visited Nice, and then, he writes to Underwood, 'I crossed the Alps by the Col de Tende, stayed at Turin three days, and came here [Genoa] through snow and ice over the Bochetta, where I have been waiting for a fair wind for Tuscany. We have had no impediments except from the snow and the east winds.' It is from Faraday's journal that we realize what the snow and the east wind meant in adding to the rigours of the way, and to the difficulties of conveying a carriage and quantities of baggage over the mountains in winter. Sixty-five men with sledges and a number of mules were employed. Faraday's entry in his journal conveys, without inflated sentiment, the wild route, the movements of the men, and his own feelings. He never tries by his writing to make a thing sound more interesting than it is; he is exactly the opposite to a romantic in his reaction to desolate places and towering peaks; he is 'sense' contrasted with Jane Davy's 'sensibility' and Davy's genuine elation in the presence of the sublime. Faraday writes like a pre-romantic, 'The view from the elevation was very peculiar and if immensity bestows grandeur was very grand.' At Genoa they waited in vain for a fair wind; Faraday writes that they had a very disagreeable passage in an open boat to Lerici; and were not without apprehensions of being overset on the way.

By March 16th Davy was in Florence. As early as 1808 he had made experiments on the nature of charcoal and the diamond; at Florence he used the great lens in the laboratory of the Accademia del Cimento to make some experiments on the combustion of the diamond and some other carbonaceous substances, from which he inferred that they were chemically the same, thus overthrowing the generally held belief 'that bodies cannot be the same in composition, or chemical nature, and yet totally different in all their physical properties'. Later, in Rome, in the laboratory of the Accademia Lincei, Davy continued his researches on the combustion of different varieties of carbon.

Berzelius named these bodies, having different physical properties though alike in their chemical nature, allotropes. In communicating his results to the Royal Society, Davy named the scientists by whose assistance he had been honoured in Florence and Rome.

Of the Italian scientists with whom Davy worked, Morichini was the one with whom he formed the strongest personal friendship. He was happy in Rome, which he reached early in April. Again and again, both in his notebooks and in his last piece of writing, *Consolations in Travel*, he reveals how profound was the impression made on him by Rome and the Roman environs. Jane, too, was happy in Rome as, perhaps, nowhere else. Davy still worked. But even with him work must have taken a secondary place that April; even Faraday was infected with excitement. Napoleon had abdicated. Faraday wrote a letter to his mother from Rome, dated April 14th, 1814. 'We are at present in a land of friends, and where every means is used to render communication with England open and unobstructed. Nevertheless this letter will not come by the ordinary route, but by the high favour of Sir H. Davy will be put in with his own, and it will be conveyed by a peculiar person.' In a postscript he added, 'We have heard this morning that Paris was taken by the allied troops on March 31, and, as things are, we may hope for peace, but at present all things are uncertain. Englishmen are here respected almost to adoration, and I proudly own myself as belonging to that nation which holds so high a place in the scale of European powers.'

On May 1st, in a letter to Abbot, Faraday again described the favour with which his countrymen were everywhere received. He wrote:

Since we have left the French dominions, we have been received with testimonies of joy and gratitude as strong as it was possible for the tongue to express. At Lucca, we found the whole population without the gates, waiting for the English. It was said that the army which had embarked at Leghorn would enter Lucca that day, and the inhabitants had come out to receive them as brothers. The town was decorated in the most brilliant manner by colours, drapery, and embroidery flying from every window, and in the evening general illuminations took place, done as expressions of their joy at their deliverance from the French government; and the English were hailed everywhere as their saviours.

The First Continental Journey

Months later, the English were still so much in favour that in one of the passes of the Apennines, between Rome and Naples, Davy was allowed to go unmolested by the captain of a group of bandits in compliment to his country. Davy told his brother John that he had a conversation with the bandit in walking up the steep ascent of the road beyond Fondii. On another occasion, still later in the course of their journeying, an incident happened which was often repeated by Sir Walter Scott. In 1831 Scott told the story to John Davy's wife and John's wife wrote it down immediately as nearly as possible in Scott's words. He said he had formed Humphry's acquaintance when he was young in the voyage of life, and happy in the enjoyment of simple pleasures. He then continued:

There was one very good thing about him, he never forgot a friend; and I'll tell you a thing he did for me which makes me particularly say so. When he was travelling in the Tyrol, the old patriot Speckbacker was very ill, suffering from rheumatism, or something of that sort; and when he heard that there was a great philosopher in the neighbourhood, he thought of course he must be a doctor, and sent to beg some advice about his complaint. Sir Humphry did not profess to know much about medicine, but he gave him something which luckily relieved the pain; and then the gratitude of the old chief made him feel unhappy because he refused any fee. So Sir Humphry said, 'Well, that you may not feel unhappy about not making me any return for my advice, I'll ask if you have any old pistol, or rusty bit of a sword that was used in your Tyrolese war of defence, for I have a friend who would be delighted to have any such article; and you may depend on its being hung up in his hall, and the story of it told for many a year to come.' Speckbacker struck his hands together much pleased at this request, and said, 'Oh, I have the very thing! You shall have the gun that I used myself when I shot thirty Bavarians in one day.' The illustrious gun was given accordingly to Sir Humphry, who brought it with him on his next visit to Scotland, and deposited with me at Abbotsford himself.

While the Davys were in Rome, on May 24th Pope Pius VII (Pope from 1800 to 1823) was rejoicingly received back in Rome from his long exile. He had been a prisoner in Grenoble, Savona, and Fontainebleau; he had striven to maintain the independence of his

See, had refused to join an offensive alliance against England, and, harried and desolate, had shown a courage which had made his release one of the cementing aims of the allies. John Davy thinks it was at Rome, and not, as sometimes stated, at Fontainebleau, that Davy paid his respects to a Pope 'whose sanctity, firmness, meekness and benevolence, he considered an honour to his church and human nature'. Humphry witnessed the Pope's re-entry into Rome; saw him borne on the shoulders of the most distinguished artists headed by Canova, and wrote of the emotion shown by every section of the people. In later years he was to regret the 'domestic policy of extreme reaction' followed by the successor of Pius VII – Leo XII. In *Consolations in Travel* he makes more detailed reference to the return of the Pope; but he is writing then in an assumed character and of a transferred incident.

From Rome Davy went to Naples and saw, at last, Vesuvius. In 1808 he had begun to entertain the idea that volcanic eruptions were due to chemical agencies; that there was no other adequate source for volcanic fires than the oxidation of the metals which form the bases of the earths and alkalis. He cherished this opinion for twenty years, returning to Vesuvius, and making in his laboratory many, and some dangerous, experiments in the hope of confirming his hypothesis. Only at the very last, in the Third Dialogue of *Consolations in Travel*, did he surrender his hypothesis, when he made the Stranger conclude that volcanoes were owing to a central fire, this notion being more consistent with the known facts, and 'more agreeable to the analogy of things'.

He made observations on Vesuvius in the spring of 1814 and again in the spring of 1815; in December 1819 and in January and February 1820. It was during the 1819–20 visit that the volcano offered the most favourable opportunity for investigation, and it is this account which is most vivid in the paper *On the Phenomena of Volcanoes* read before the Royal Society in 1828. But he also used observations made in May 1814, and his more general memories of the Naples scene remained sovereign in his recollection, in spite of the negative result of his researches. At this time he paid his first visit to Paestum.

In June 1813 the travellers visited Bologna, Mantua, Verona, and Milan, and it seems to have been during this June that Davy spent some time in Pavia with Volta, whom he considered the greatest living light in physical science.

The First Continental Journey

At Pavia it was Davy's European reputation as a scientist which enabled him to exchange views with so illustrious a scientist as Volta; in Geneva, to which the travellers went by the Simplon Pass to avoid the Italian summer heat, it was Jane Davy's friendship with Madame de Staël which brought him into relation with Madame de Staël herself and her party. While at Geneva the Davys had a villa about three miles from the city itself; Davy could angle from the garden which ran down to the water's edge. He could talk with those who were his superiors in fields not his own, as well as with De la Rive and other scientists of Geneva. Jane was happy in Geneva as in Rome. After three months spent in Geneva, Davy travelled in the Austrian Tyrol amidst scenery which, for grandeur and variety, he came to care for above all other.

The winter, like the preceding spring, was spent chiefly in Rome and Naples. Davy continued his work on volcanoes and he also continued an examination of the nature and composition of the colours used by the ancient Greeks and Romans, with experimental attempts to imitate such pigments as were peculiar. His experiments were made on colours found in the frescoes of the baths of Titus, in the ruins known as the Baths of Livia, in the remains of other palaces and baths of Rome and in Pompeii. Canova, a true friend of Davy, who at that time was charged with the care of works connected with ancient art in Rome, enabled him to select with his own hands specimens of the different pigments. From Murat, King of Naples when Davy had first visited it, he had also received much courtesy. Murat was, said Davy, a patron of the fine arts: 'more was done in his short and feverish reign in bringing to light the treasures of antiquity buried in Pompeii than before or since in a triple span of time'. Davy's paper *On the Colours used in Painting by the ancient Greeks and Romans* was read before the Royal Society on February 23rd, 1815. It is reprinted from the *Philosophical Transactions* of the same year in the sixth volume of the collected *Works*. These valuable analyses remain interesting even to those not technically concerned with cinnabar and cerulean, Tyrian purple, orpiment, and sandarach.

Occupied with the varied work he loved, mingling with an equally varied society, and exercising his gift for happy speculation, Davy expanded; but Faraday was often in a state of dejection, living, as it were, in no-man's-land. At Geneva, De la Rive had appreciated and encouraged him, and the two maintained afterwards an interesting correspondence. The Genevan professor is said to have remarked that if he could not have Faraday to dinner with Sir Humphry and Lady Davy he would give a separate party for him. As 'philosophical assistant' to Davy, Faraday recognized to the full the tremendous advantages he enjoyed; 'the constant presence of Sir H. D. is a mine of inexhaustible knowledge and improvement,' he wrote to Abbot, 'and the glorious opportunities I enjoy of improving in the knowledge of chemistry and the sciences continually determine me to finish the voyage with Sir H. D.' He added, however, 'But if I wish to enjoy these advantages I have to sacrifice much, and though these sacrifices are such as an humble man would not feel, yet I cannot quietly make them'. Faraday said several times that it was not Sir Humphry who mortified him; on the contrary, Sir Humphry seems to have done his utmost not to hurt Faraday by requiring of him personal duties. It was his habit to look after himself. Faraday admits it. He also warned Abbot after voicing his complaints, 'I wrote in a ruffled state of mind which, by the bye, resulted from a mere trifle'. Possibly people in Rome treated him differently from the way he had been treated at Geneva, for it was during the second visit to Rome, after the visit to Geneva, that the letters of grievance were written. Like all people voicing a grievance, his breast within him swelled as he wrote, and he went back to the beginning of the matter, the beginning of all the trouble being that Davy never kept his promise to secure a personal assistant who would act instead of Faraday as intermediary with the servants in making arrangements and in managing expenses. On the other hand, Faraday never learnt enough French or Italian to be of much use with such a servant could he have been found. Faraday set out the circumstances minutely both to Abbot in November 1814 and to Abbot and Huxtable in the January and February of 1815. He was lonely, and unable to adapt himself to the countries through which he passed; he was cut off from his religious associates and felt a deep antipathy to the Catholic Church as he beheld it; he did not feel that he could be religious in Paris and Rome; and always there was the quick fear of being thought a lackey.

A good deal of nonsense has been written about the wrongs of Faraday during this first continental journey. Faraday was a genius but he had not, up to that time, shown it. By natural endowment he was entitled to far greater dignity than he assumed; but Jane Davy treated him according to her lights. Accustomed as she was to European society, and to the attitude of the London and Edinburgh *savants* towards the French, her own 'treasure', with his anti-gallican feelings, must often have seemed a bore; and his protégé insufferable. Faraday forgave her. Hers are said to have been among the few evening parties he attended when, in the course of time, he had made his classic discoveries and, in his turn, become a famous man. After all, both philosophers owed much to the lady. Without her they would have had to walk. Instead Faraday wrote, 'We seem tied to no spot, confined by no circumstances, at all hours, at all seasons, and in all places we move with freedom – our world seems extending and our existence enlarged; we seem to fly over the globe, rather like satellites to it than parts of it, and mentally take possession of every spot we go over.'

Davy had intended, after his stay in Rome and Naples, to extend his journey to Greece and the Near East; 'it is a moral certainty that we are to see Constantinople' Faraday had written to his mother. But because of a fresh outbreak of the plague, and because Davy refused to consider undergoing the quarantine regulations, this project was abandoned; and the return of the party was speeded by the news of Napoleon's escape from Elba. Faraday shows his absorption in improving his scientific knowledge to the exclusion of everything else by the entry in his diary: 'March 7th, 1815, I heard for news that Bonaparte was again at liberty. Being no politician I did not trouble myself much about it, though I suppose it will have a strong influence on the affairs of Europe.' But his excitement at the prospect of returning home warms our hearts to him, and his letter gives a swift impression of their course and of their speed at that time. He wrote from Brussels on April 16th, 1815.

My very dear Mother – It is with no small pleasure I write you my last letter from a foreign country . . .
I am not acquainted with the reason for our sudden return; it is however sufficient for me that it has taken place. We left Naples very hastily, perhaps because of the motion of Neapolitan troops, and perhaps for private reasons. We came rapidly to Rome, we as rapidly left it. We ran up Italy, we crossed the Tyrol, we stepped over Germany, we entered Holland, and we are now in Brussels, and talk of leaving it tomorrow for Ostend; at Ostend we embark, and at Deal we land on a spot of earth which I will never leave

again. You may be sure we shall not creep from Deal to London, and I am sure I shall not creep to 18 Weymouth Street; and then – but it is no use, I have a thousand times endeavoured to fancy a meeting with you and my relations and friends, and I am sure I have as often failed, and the reality must be a pleasure not to be imagined or described . . . I shall be thankful if you will make no inquiries for me anywhere, and especially not in Portland Place, or of Mr Brande. I do not wish to give occasion for any comment whatever on me or mine.

Davy reached London on April 23rd to learn that John Davy, now fully qualified as a doctor, had decided to join the medical department of the army, a disappointment to his mother and sisters who had hoped that when Doctor Ayrton Paris (who afterwards wrote Davy's biography) moved from Penzance, John might succeed him. Humphry wrote to his youngest sister, Betsy, in relation to this plan:

June 1, 1815.
My dear Betsy,

I have received your kind letter. It is not improbable that had your letter arrived before John's plans were fixed that he would have hesitated; but he was at Ostend before I received it; and now it is too late to change his views. Should he not like the service, however, the situation at Penzance is a resource; and it will always be in his favour, as a medical man, to have seen practice in the army. It is, indeed, very possible that the campaign may be finished, and he on his return long before Dr Paris gives up Penzance, and if so, it will be a subject for consideration. For my own part, I do not think that Buonaparte and his diabolical adherents can long oppose the allies; and the French are too selfish to adhere to him for any other reason than the hope of plundering their neighbours. John had a tedious passage, but landed quite well, and writes to me in excellent spirits.

I write this letter to inform you of his safe arrival on the continent. I am, with kindest love to my mother and sisters,

Your ever affectionate brother,

H. DAVY.

It must have seemed hard to the girls that, no sooner had one of their brothers returned from the Continent than the other must set out

[158]

for it. But Humphry's expectation of the speedy defeat of Bonaparte, doubtless expressed with special confidence so as to allay Betsy's fears, was justified; less than three weeks later the Duke of Wellington was writing to his brother, William Wellesley-Pole:[1]

Bruxelles
June 19th, 1815.

My dear William,

You'll see the account of our desperate Battle and victory over Boney!!

It was the most desperate business I ever was in; I never took so much trouble about any Battle; and never was so near being beat.

Our loss is immense particularly in that best of all Instruments – British Infantry. I never saw the Infantry behave so well.

I am going immediately.

Can we be reinforced in Cavalry or Infantry or both?

We must have Lord Combermere as Lord Uxbridge has lost his leg. He was wounded when talking to me during the last attack, almost by the last shot.

Ever your most affectionately

w.

Before July was out the warship *Bellerophon*, under the command of Captain Maitland, had sailed into Plymouth Sound with Napoleon aboard on his way to his final exile. A description of him by Davy's fellow-townsman, the Rev. Charles Valentine Le Grice,[2] perpetual curate of St Mary's, Penzance, in a letter to Mr Samuel John, a solicitor at Penzance, makes a fitting appendage to Faraday's description of him in Paris. Le Grice's letter was posted at Stonehouse in August 1815. It runs:

Fore-street, Stonehouse, Tuesday – My dear Friend, the mail coach was empty, the tide when we arrived at Torpoint favourable and at ¼ past six yesterday evening, I saw standing alive under the

[1] *Some Letters of the Duke of Wellington to his Brother William Wellesley-Pole*, ed. Professor Sir Charles Webster, Camden Miscellany, vol. xviii, p. 35.

[2] Published for the first time in the *Western Morning News*, July 30th, 1955, and reprinted here by permission of the Editor.

protection of our glorious flag amid thousands of spectators, so thick together that the sea appeared literally like dry land – Napoleon Buonaparte.

The rush of ideas that overwhelmed my mind at the moment he appeared on the side of the ship can be conceived, but not enumerated or delineated.

He stood uncovered in full dress, on the step of the gangway, his Marshals in a compact circle behind him all uncovered, certainly a most majestic figure. He stood there ten minutes then bowed, and retired. We again saw him at the cabin window very plain.

He is very stout, firm, and athletic, inclined to corpulence, but the prominence of his belly is partly owing to the shape of his coat, which is not more than six inches deep on the breast, and then slopes off, more like the lapels of a coat, than a coat. He has the handsomest legs I ever saw. He is like the profile on my box, and on his large silver coin.

The scene itself was interesting, the sea covered with boats and the breakwater with hundreds of people standing upon it, it being then low water, but I could look at nothing but him, whose Bust must fill so conspicuous a niche in the History of the World . . .

As to the probability of your seeing him, all I can say is . . . use your own discretion. The people here know nothing. Buonaparte appears regularly at half-past five or six o'clock in deck, but so little is known, that Lord Mount Edgcombe and a party rowed off understanding he would not appear, and in about half an hour he came on deck . . .

The *Bellerophon* lies within the Breakwater about a mile, which at sea seems a short distance. Buonaparte was dressed in blue coat, red cuffs, white breeches and white silk stockings, his hat in one hand, his glass in the other.

Valentine Le Grice is representative of those Englishmen who saw Napoleon in a heroic and romantic light. Davy brooded only on his annihilating egoism. Had he been as good a poet he might (remembering the retreat from Moscow) have said with Walter De La Mare in our own day:[1]

---

[1] 'Napoleon', from *Poems 1901–1918*, Constable & Co., London, 1920.

What is the world, O soldiers?
     It is I:
I, this incessant snow,
     This northern sky;
Soldiers, this solitude
     Through which we go
     Is I.

13 · *The Safety Lamp*

The war was over. Humphry and Jane bought a house in Grosvenor Street and prepared to settle down. John Davy did not return to England either to become a physician in Penzance or to make, as perhaps his brother had hoped, research his career. He remained in the services, doing good work, seeing the world, and rising steadily to eminence in his profession.

Michael Faraday, a fortnight after his return to London, was offered a position in the Royal Institution a little similar to that which Edmund Davy had held before going to Dublin. Michael in his turn was given a room of his own at the Royal Institution and provided with candles. He received, as assistant in the laboratory, and as superintendent of the mineralogical collection and of the apparatus, a salary of thirty shillings a week. Within six months he was giving his first course of lectures at the City Philosophical Society; and within a year his first paper, an analysis of native caustic lime, was published. Forty years afterwards he wrote, 'Sir Humphry Davy gave me the analysis to make as a first attempt in chemistry, at a time when my fear was greater than my confidence, and both far greater than my knowledge; at a time also when I had no thought of ever writing an original paper on science. The addition of his [Sir H. D.'s] own comments, on the publication of the paper, encouraged me to go on making, from time to time, other slight communications . . .'

It was with Faraday sometimes as his assistant in experiments that Davy engaged in the investigations on flame which led to the invention of the safety lamp.

As coal-mining went deeper, disasters had come to seem so irremediably inherent, in an age when the role of air in combustion had only just become understood, that people appeared resigned to dreadful happenings. Mining was not, in the same way as negro slavery, a human institution, to be relieved by human and political action. Thomas Young and Tom Poole were among those who went without sugar as part of the widespread protest against negro slavery; but I cannot find that they refused to burn coal as a protest against loss of life among colliers. It was a Mr Wilkinson, a London barrister, who suggested the establishment of a society for spreading knowledge of the fearful accidents in fiery collieries, and for seeking by every manner of means some method of prevention, either by ventilation, or by the invention of a light which would be safe in explosive mixtures.

The immediate cause of the formation of the Society was an explosion at Felling Colliery, near Sunderland, which happened on May 25th, 1812. The loss of life and the hideous suffering entailed were brought home to people beyond the immediate neighbourhood by a narrative of the disaster which the Reverend John Hodgson published as an introductory document to the funeral sermon he preached when the number of the dead was known to be ninety-two pitmen. The narrative was reprinted by Ayrton Paris in his biography of Davy; it throws light on the nature of the evil which Davy combated, on the disaster itself, and on the conflicting passions roused. A sentence like 'Their progress [of the rescuers] was very soon intercepted by the prevalence of choak-damp, and the sparks from their steel-mills fell into it like dark drops of blood' tellingly conveys the inadequacy of existing uses and the extreme necessity for speed in the invention of a better provision for lighting. Hodgson began his narrative of the disaster at the Felling Colliery by showing how the terror of accidents was intensified by the impossibility of anticipating danger. And Felling Colliery was considered a model in its arrangements.

The Association suggested by Mr Wilkinson was formed not only to make people conscious of the horror of accidents in mines but to invite scientific men to make suggestions likely to lead to more secure methods of lighting. The Society was established on October 1st, 1813, by Sir Ralph Milbanke, afterwards Sir Ralph Noel (Byron's father-in-law), Dr Gray, Dr Pemberton, Mr Robinson, Mr Stephenson, and other gentlemen; and was under the patronage of, among others, the Bishop of Durham and the Duke of Northumberland. Some effort was made to get into touch with Davy in 1813, but he was already in France. It was not until August 1815 that his aid was definitely invited. It is, I think, in view of subsequent events, worth noting that by then two years had elapsed since the formation of the Society, and more than three since the accident in Felling Colliery, during which time empirical methods by practical engineers had had ample time to establish themselves successfully had success been within their immediate scope. Dr Gray wrote to Davy at his London address and received the following reply from him at Lord Somerville's, near Melrose:

To the Reverend Dr Gray

Aug. 3, 1815.

Sir,

I had the honour of receiving the letter which you addressed to me in London, at this place, and I am much obliged to you for calling my attention to so important a subject.

It will give me great pleasure if my chemical knowledge can be of any use in an inquiry so interesting to humanity, and I beg you will assure the Committee of my readiness to co-operate with them in any experiments or investigations on the subject. If you think my visiting the mines can be of any use, I will cheerfully do so.

There appear to me to be several modes of destroying the fire-damp without danger; but the difficulty is to ascertain when it is present, without introducing lights which may inflame it. I have thought of two species of lights which have no power of inflaming the gas which is the cause of fire-damp, but I have not here the means of ascertaining whether they will be sufficiently luminous to enable the workmen to carry on their business. They can be easily procured, and at a cheaper rate than candles.

I do not recollect anything of Mr Ryan's plan; it is possible that it has been mentioned to me in general conversation, and that I have forgotten it. If it has been communicated to me in any other way, it has made no impression on my memory.

I shall be here for ten days longer, and on my return South, will visit any place you will be kind enough to point out to me, where I may be able to acquire information on the subject of the coal gas.

Should the Bishop of Durham be at Aucland, I shall pay my respects to his Lordship on my return.

I have the honour to be, dear Sir, with much respect, your obedient humble servant,

H. DAVY.

The Safety Lamp

About a fortnight later he wrote again to Dr Gray in reply to a letter received:

Melrose, August 18, 1815.

Sir,

I received your letter, which followed me to the Moors, where I have been shooting with Lord Somerville. I should have replied to it before this time, but we were in a part of the Highlands where there was no post. I am very grateful to you for the obliging invitation it contains.

I propose to leave the Tweed-side on Tuesday or Wednesday, so I shall be at Newcastle either on Wednesday or Thursday. If you will have the kindness to inform me by a letter, addressed at the Post Office, where I can find the gentleman you mention, I will call upon him, and do anything in my power to assist the investigation in that neighbourhood.

I regret that I cannot say positively whether I shall be at Newcastle on Wednesday or Thursday; for I have some business at Kelso which may detain me for a night, or it may be finished immediately.

I am travelling as a bachelor, and will do myself the honour of paying my respects to you at Bishop-Wearmouth towards the end of the week.

I am, Sir, with much respect,
Your obedient humble servant,

H. DAVY.

The gentleman whom Dr Gray had mentioned was Mr John Buddle, a practical engineer in charge of the Walls-End Colliery, a man of great ability and authority in all the operations of coal-mining, a man of determination, courage, and loyalty. His account of Davy's visits to Walls-End Colliery and to Morden West, one of the mines belonging to the Earl of Durham, together with his letters and detailed testimony, need to be read as a supplement to Davy's own exposition of his inquiry in *On the Safety Lamp; With some Researches on Flame*, reprinted with various interesting extracts and observations in the sixth volume of Davy's collected *Works*. Some of John Buddle's statements are there reprinted. But all John Buddle's evidence, given years later, is worth reading. It is printed in the *Report from the Select Committee on Accidents in Mines*, published on September 4th, 1835.

He emerges as a forthright, emphatic man. It is well that Davy liked him living; for Buddle was to be his most stalwart supporter when Davy was dead and could no longer speak for himself. It is a proof of something very genuine in Davy that Buddle not only paid homage to his genius but liked him as a friend. In later life he always visited Davy when he was in town and Davy returned the visits when he was in the North.

Davy knew little about coal-mining; but he was well informed about another kind of mining and, as a Cornishman, and a Cornishman brought up in the vicinity of tin and copper mines, he had a natural feeling for a mining community. Accompanied by the Rev. John Hudson, he first visited the Walls-End Colliery on August 24th, 1815, and John Buddle, who had already been introduced to Davy by Dr Gray, explained as far as he was able the nature of fiery coal-mines.

Buddle and Davy conversed for a very long time, the one having detailed practical knowledge, the other a mind stored with scientific learning which reached out in many directions. Buddle considered that the resources of modern chemical science had been fully applied in his own improved plans of ventilation. The comparative lightness of fire-damp was well understood; every precaution was taken to preserve the communications open; and the currents of air were promoted or occasioned, not only by furnaces but likewise by air-pumps and steam apparatus. The great desideratum, Buddle said, was not further ingenuity in ventilation but a light that could be safely used in an explosive mixture. He explained to Davy the degree of light necessary for working underground. He had, he said, at that time, not the slightest idea of ever seeing accomplished a safe lamp – so long and ardently sought.

He wrote, 'After a great deal of conversation with Sir Humphry Davy, and he making himself perfectly acquainted with the nature of our mines, and what was wanted, just as we were parting he looked at me and said, "I think I can do something for you." Thinking it was too much ever to be achieved, I gave him a look of incredulity; at the moment it was beyond my comprehension. However, smiling, he said "Do not despair, I think I can do something for you in a very short time." '

The tenor of Davy's days had equipped him singularly well for the investigation suggested. He had a mind prodigal in expedients; and his native dash and courage speeded his knowledge. Yet it is the gradual progress of the discovery which is so beautifully revealed in his successive papers. He began with a minute chemical examination of various specimens of fire-damp sent from the North to him and to his friend, Children; he made numerous experiments on the circumstances under which it exploded and the degree of its inflammability; he shows by what train of reasoning he reached the idea of flame-sieves or apertures in the construction of a lamp; and by what complicated combinations he arrived finally at the simple principle of surrounding the light entirely by wire-gauze, and making the same tissue feed the flame with air and emit light. Then he writes, 'In plunging a light surrounded by a cylinder of fine wire-gauze onto an explosive mixture I saw the whole cylinder become quietly and gradually filled with flame; the upper part of it soon appeared red hot; yet *no* explosion occurred.' In the earliest lamps he devised, the air-feeders were such that, although they gave protection to the user against explosion, their light was extinguished in explosive mixtures in which the fire-damp was in sufficient quantities to absorb the whole oxygen of the air. The wire-gauze lamps, which were ready by January 1816, not only gave protection but acted as detectors, for they burned more brightly at the approach of danger, rising to give light, the very element whose properties had formerly caused destruction.

John Buddle and Dr Hodgson were the first who, confident in the strength of Davy's authority, took a safety lamp into a part of a mine where there was fire-damp. Davy had warned Buddle that there was no hazard except in exposing the lamp to a strong current; he cautioned Buddle against such exposure, but said he hoped he had a remedy for it. Buddle writes that he first tried the lamp in an explosive mixture on the surface, and then took it into a mine. It was impossible, he said, to express his feelings of astonishment and delight when he first suspended the lamp in the mine and saw it red hot; he said that if a monster had been destroyed he could not have felt more exultant. He exclaimed to those around him, 'We have subdued this monster.'

Davy's own papers must be read by anyone who wishes to realize both the value of the original discovery, the practical application, the quick improvements Davy effected, and the careful pains he took to try to guard careless users of his lamps against themselves. He went to the North to demonstrate the precautions necessary to be taken in using the lamps in the only conditions in which they could explode –

i.e. in a strong current of gas or strong current of explosive mixture, when there was a risk of the passing of the flame through the gauze. John Buddle described this visit and another paid some time after the lamps had been in use at Walls-End.[1] When asked if he could define the time in hours and minutes during which they remained underground Buddle said:

No, I cannot, but I should suppose from a quarter of an hour to half-an-hour, during which Sir Humphry was explaining to us the nature of the lamp; it was a lecture in fact upon the spot, and how the time might pass during that interval I cannot tell, but I think we might be down the mine for a period of two or three hours altogether. He then explained to us the danger of exposing the lamp to a strong current of gas, or even to a strong current of explosive mixture, as it would risk the passing of the flame through the gauze, but he pointed out a remedy at the same time for that contingency, and which we have always used; namely, by a tin screen, which slides upon the frame-wires of the lamp, and encircles the circumference of the gauze cylinder to an extent of about one-half to two-thirds of its circumference. He could not there show us the effect of the passing of the explosion to make us sensible of the danger to which we were exposed in that way, and teach us how to avoid it. But some time previously to that I accompanied him to one of the Earl of Durham's collieries, – to what is called the Morton West pit, where a very large blower from the shaft, not from the coal but from a fissure in the stone, had been for many years discharging up the shaft, by a cast-iron pipe to the surface in a similar manner to that which I have already described at Walls-End. We took a length of hose from an extinguishing engine, with the jet-pipe upon it, and attached that to the blower-pipe at the top of the pit; it was held horizontally, and the jet was thrown very forcibly out of the nozzle of the pipe; the blower was sufficiently strong to propel the stream of gas across the engine-house. I well recollect the pipe was held at the entrance of the engine-house, and the jet passed the explosion nearly to the far end of the room, for it was very powerful; the distance that the blower fired it was from nine to twelve feet I should think. I held the lamp in the direction

[1] Buddle was speaking before a Select Committee in 1835, after another accident had occurred in this pit.

of the jet, and not having seen it before, I was not very apprehensive of its firing. It did not fire at first, but as I approached the end of the nozzle-pipe, the gauze became heated red-hot and passed the explosion. The flame was as long or longer than the breadth of the engine-room; I remember that it burnt the nap off my great coat and spoiled it. This experiment was repeated over and over again. Lord Durham himself was present, and a great many other persons, professional men and others, were present on this occasion. The force of Sir Humphry's remarks at the time was 'Now, gentlemen, you see the nature of the danger to which you are exposed in using the lamp, and I caution you to guard against it in the manner I have shown you. This is to show the only case in which the lamp will explode, and I caution and warn you not to use it in such case when you can avoid it without using the shield.'

In addition to this practical demonstration of the need for care made before those responsible for certain collieries, Davy wrote much on the safety lamp and on precautionary measures expressly designed for practical purposes; but though published in a cheap form it was in so little request that a second edition was never required. We are, as all history shows, invincibly careless creatures. We take a chance. We hate trouble. Smoking is pleasant. Any open flame is for many purposes more useful than a flame imprisoned. Goldsworthy Gurney, giving evidence before the Select Committee in 1835, said, 'I believe miners will open their lamps; it is said they pick them when locked.'

The sequel to the invention of the safety lamp is an ironical comment on human behaviour and the pity of it. There was magnanimity. Davy, never a chafferer, gave his invention freely and, thinking to benefit humanity, refused to take out a patent, though urged to do so by John Buddle and others. His lamps were extensively introduced into the collieries of Tyne and Wear, and on September 13th, 1817, a group of coal-owners of the Tyne and Wear, magnanimous in their turn, presented Davy, in token of their high appreciation, with a valuable service of plate. The occasion of the presentation was one on which they sought to do their benefactor every honour. A dinner was held in the Queen's Head at Newcastle, attended by the great proprietors of collieries, representatives of the coal trade, by those gentlemen who had suggested the investigation and those who had helped to organize the use of the lamp in the mines. It must have been a moving occasion for Davy. Mr J. G. Lambton (first Earl of Durham), who was in the chair, and who, as the festivities drew to a close, addressed Davy before making the presentation, was once the boy Davy had known when he lived with Dr Beddoes. Perhaps it is not altogether fanciful to see in the address itself something of Beddoes; for Lambton retained the little Doctor's principles, loved to put them into action, was careful of his workmen, and highly popular with them. Lambton said:

Sir Humphry – It now becomes my duty to fulfil the object of the meeting, in presenting to you this service of plate, from the Coal-owners of the Tyne and Wear, as a testimony of their gratitude for the services you have rendered to them and to humanity.

Your brilliant genius, which has been so long employed in an unparalleled manner, in extending the boundaries of chemical knowledge, never accomplished a higher object, nor obtained a nobler triumph.

You had to contend with an element of destruction which seemed uncontrollable by human power; which not only rendered the property of the coal-owner insecure, but kept him in perpetual alarm for the safety of the intrepid miner in his service, and often exhibited to him the most appalling scenes of death, and heart-sickening misery.

You have increased the value of an important branch of productive industry; and, what is of infinitely greater importance you have contributed to the safety of the lives and persons of multitudes of your fellow-creatures.

It is now nearly two years that your Safety Lamp has been used by hundreds of miners in the most dangerous recesses of the earth, and under the most trying circumstances. Not a single failure has occurred – its absolute security is demonstrated. I have, indeed, deeply to lament more than one catastrophe, produced by fool-hardiness and ignorance, in neglecting to use the safe-guard you have supplied; but these dreadful accidents even, if possible, exalt its importance.

If your fame had needed anything to make it immortal, this discovery alone would have carried it down to future ages, and connected it with benefits and blessings.

Receive, Sir Humphry, this permanent memorial of our profound respect and high admiration – a testimony we trust, equally honourable to you and to us. We hope you will have as much pleasure in receiving, as we feel in offering it. Long may you live to use it – long may you live to pursue your splendid career of scientific discovery, and to give new claims to the gratitude and praise of the world.

This speech by Lambton sums up what Davy most valued in the praise bestowed on him for his invention; it gave him recognition and gratitude. As the value of his discovery became known, and its effects were utilized not only in coal-mines but for preventing explosions in spirit houses and magazines in ships, he reached a pinnacle of honour in the eyes of ordinary people as well as in the eyes of those highly placed, hardly paralleled by any other man of science in his day. The lamps began to be used, and called 'Davys', wherever, in England or Europe, fire-damp was a menace. Nor was Davy's praise confined to practical men; the discovery won him the praise of fellow-scientists. The warmest letter Sir Joseph Banks ever wrote him was dated October 30th, 1815, after he had received Davy's first communication to the Royal Society on the Safety Lamp for preventing explosions in mines. Davy's letter, he said, had given him unspeakable pleasure. 'Much as, by the more brilliant discoveries you have made, the reputation of the Royal Society has been exalted in the scientific world, I am of opinion that the solid and effective reputation of that body will be more advanced among our contemporaries of all ranks by your present discovery, than it has been by all the rest.' Later, for his extended researches on flame, the President and Council of the Royal Society bestowed on him the Rumford Medal. In 1818 the Prince Regent created him baronet. Nor was his triumph short-lived. As late as 1825 he received from the Emperor Alexander of Russia 'a superb silver gilt vase standing in a circular tray enriched with medallions. On the cover was a figure, of about sixteen or eighteen inches in height, representing the God of Fire weeping over his extinguished torch.' This vase is now in the keeping of the Royal Institution. And because there is something in the story of the lamp which catches the imagination of children – like all tales of fiery dragons disarmed by a trick – Davy has won the kind of immortality which he, the lover of Aladdin, would not have despised.

But side by side with all this laud and honour went an accusation which embittered all his pleasure. Davy had begun his experimental investigation in October, 1815; by the following January his first lamps were being introduced into the mines. Although he had worked as it were in full daylight, revealing his processes of thought, his gropings towards attainment, and his successive improvements as he went along, he was accused of having filched his original idea from Dr Clanny and George Stephenson. The counter-claim for Stephenson was voiced when Davy was in the flush of success, and was entertained partly because of that element in all discovery which is 'alert recognition', and partly because Stephenson was as great a mechanical as Davy a scientific genius. Claims as to priority of discovery were made by Stephenson's friends rather than by himself. He was working on mechanical principles and certainly had not produced a safe lamp. Whether, as is often claimed, he might have done so by a process of trial and alteration is, after all, a matter of conjecture. He might as possibly have blown himself up. Davy recognized that wire gauze consists of a very large number of very short fine parallel tubes and brought a lifetime's scientific interest in flame to bear upon a practical problem.

Davy was never a meek man. Various letters and speeches reveal the intensity of his anger at the claims made for Stephenson. Sometimes he was heavily sarcastic, saying that he had no desire to go out of his way to crush gnats that buzzed in the distance and did not bite him. But in a letter to John Buddle – a letter recently published in a Paper by Dr E. Weil, in *Annals of Science*, vol. vi, no. 3, Feb. 1950, he gives a forthright expression of wrath. The letter was dated February 13th, 1817, and was mainly concerned with details of further suggested improvements to the lamp. It was sent to John Buddle at Walls-End:

. . . I was very unwell when I received the Newcastle paper containing these infamous advertisements & they made me more angry than they ought to have done.

I wish Stevenson & his attempts through Brandling had been always greeted with silent contempt. Give a knave rope sufficient & he will hang himself & he has I think effectually done so in his miserable pilfering lying & equivocating pamphlet.

I shall in the Course of the Spring, publish a work on Flame & on the principles of the Lamp. I shall be glad to attach to it some practical evidence of your pupils & a short report from Mr Hodgson. – I am now satisfied with the lamp & that I am so I owe very much to you & Mr Hodgson. – I should like to see one of the new Killing-worth lamps not from mere curiosity: but because Stevenson is so ignorant that I fear he may endanger the lives of the workmen. & it will be a slight compensation to the families of the sufferers to quote to them Brandlings opinion that he is the real inventor of the wire gauze lamp – or of the steam engine or the prism to the discovery of which he has just as much claim.

Pray remember me to Mr Hodgson – I regret that I wrote him so long a letter shewing the strong points of Stevenson's piracy. His aimable & benevolent mind & disinterested love of truth & of justice & his fear to injure even the unworthy make him fit for a more dignified employment than that of refuting the lies & mis-representations of scoundrels.

I am Dear Sir with much esteem yours very sincerely

H. DAVY.

There is no point in going once again into George Stephenson's claim, or rather the claim made for him by his partisans. The evidence has many times been fully examined and is readily available. Testimony for Davy was published and signed not only at a general meeting of coal-owners of the North, held on November 16, 1817, but also by men of science who came to a meeting held for 'considering the facts relative to the discovery of the safety-lamp'. Their resolutions read as follows:

Soho Square, November 20th 1817.

An advertisement having been inserted in the Newcastle Courant of Saturday, November 7th, 1817, purporting to contain the resolutions of 'A meeting held for the purpose of remunerating Mr George Stephenson, for the valuable services he has rendered mankind by the invention of the Safety-lamp, which is calculated for the preservation of human life in situations of the greatest danger.'

We have considered the evidence produced in various publications, by Mr Stephenson and his friends, in support of his claims, and having examined his lamps and inquired into their effects in explosive mixtures, are clearly of opinion –

First, that Mr George Stephenson *is not* the author of the discovery of the fact, that an explosion of inflammable gas will not pass through tubes and apertures of small dimensions.

Secondly, that Mr George Stephenson *was not* the first to apply

[173]

that principle to the construction of a safety-lamp, none of the lamps which he made in the year 1815 having been safe, and there being no evidence even of their having been made upon that principle.

Thirdly, that Sir Humphry Davy not only discovered, independently of all others, and without any knowledge of the unpublished experiments of the late Mr Tennant on flame, the principle of the non-communication of explosions through small apertures, but that he also has the sole merit of having first applied it to the very important purpose of a safety-lamp, which has evidently been imitated in the latest lamps of Mr George Stephenson.

(Signed) JOSEPH BANKS, F.R.S.
WM. THOMAS BRANDE
CHARLES HATCHETT
WM. HYDE WOLLASTON

J. G. Children, who had been in Davy's confidence throughout, published two papers in his support; Coleridge twice emphasized that the invention was in no way the result of chance or of practice, but of thought. Faraday in May 1829 concluded a lecture on the Safety Lamp, delivered at the Royal Institution, with the words, 'Such is the philosophical history of this most important discovery. I shall not refer to supposed claims of others to the same invention, more than to say that I was witness in our laboratory to the gradual and beautiful development of the train of thought and the experiments which produced it. The honour is Sir H. Davy's, and I do not think that this beautiful gem in the rich crown of fame which belongs to him will ever again be sullied with the unworthy breath of suspicion.'

Davy did not live to hear Faraday's tribute. He must have been pleased by the support of his famous colleagues in science and letters: but he said of all the letters he received that which made him most glad that he had laboured to be useful was one signed by a group of miners. And miners may be said to have had the last word; they called their lamps not Stephensons but Davys.

14 · *The Second Continental Journey*

A man working, when occasion called, so consumedly as Davy could not have been easy to live with; the most placid woman might have been tried, and Jane Davy was not placid.

The best picture of ease in the household is by the American scholar of Cambridge, Massachusetts, George Ticknor, of whom Ferris Greenslet wrote, 'He was a naturalist in human nature, a specialist in the species gentleman . . . he gives us the social man better than anyone else.' Ticknor had had a letter of introduction to Davy, and had breakfasted with him in his house on June 8th, 1815. Of this visit he writes:

> I breakfasted this morning with Sir Humphry Davy, of whom we have heard so much in America. He is now about thirty-three,[1] but with all the freshness and bloom of five-and-twenty, and one of the handsomest men I have seen in England. He has a great deal of vivacity, talks rapidly, though with great precision, and is so much interested in conversation that his excitement amounts to nervous impatience and keeps him in constant motion. He has just returned from Italy and delights to talk of it, – thinks it, next to England, the finest country in the world, and the society of Rome surpassed only by that of London, and says he should not die contented without going there again.
>
> It seemed singular that his taste in this should be so acute, when his professional eminence is in a province so different and so remote; but I was much more surprised when I found that the first chemist of his time was a professed angler; and that he thinks, if he were obliged to renounce fishing or philosophy, that he would find the struggle of his choice pretty severe.
>
> Lady Davy was unwell, and when I was there before she was out, so I have not yet seen the lady of whom Madame de Staël

---

[1] An underestimate of his age. He was thirty-seven.

said, that she has all Corinne's talents without her faults and extravagances.

After breakfast Sir Humphry took me to the Royal Institution where he used to lecture before he married a woman of fortune and fashion, and where he still goes every day to perform chemical experiments for purposes of research . . .

A further entry was made by Ticknor in his diary when he called again at the house in Grosvenor Street. This time he wrote:

June 15th. As her husband had invited me to do, I called this morning on Lady Davy. I found her in her parlor working on a dress, the contents of her basket strewed about the table, and looking more like home than anything since I left it. She is small, with black eyes and hair, a very pleasant face, an uncommonly sweet smile, and when she speaks, has much spirit and expression in her countenance. Her conversation is agreeable, particularly in the choice and variety of her phraseology, and has more the air of eloquence than anything I have heard before from a lady. But, then, it has something of the appearance of formality and display, which injures conversation. Her manner is gracious and elegant; and though I should not think of comparing her to Corinne, yet I think she has uncommon powers.

Ticknor called yet again on June 21st on Sir Humphry from whom he said he received much courtesy and kindness. As he was going on a continental tour Sir Humphry gave him letters of introduction to various people, among them Canova, De la Rive, and Madame de Staël. Ticknor parted with his host with real regret.

The picture of Jane's parlour, looking more homely than anything Ticknor had seen since he left his own home, is unexpected. Much more in keeping with all we know of Jane is Sydney Smith's tribute to her as queen of a party in the Grosvenor Street house. In April 1818 he wrote to her, 'The Hollands wrote with great pleasure of a dinner you gave them, and certainly you do keep *one* of the most – perhaps *the* most agreeable houses in London. Ali Pacha Luttrell, Prince of the Albanians allows this.' By May of this same year Sydney was paying one of his frequent visits to Holland House and in a letter to Jane made another reference to Luttrell, with a delicious side-shaft both at the 'Prince' and her. 'You are of an ardent mind, and

The Royal Institution, the main staircase. Davy's Nairne electro static machine (*c.* 1803) is seen on the left. The portrait of Davy (*centre*) is by an unknown artist. The statue of Faraday (*right*) is by J. H. Foley

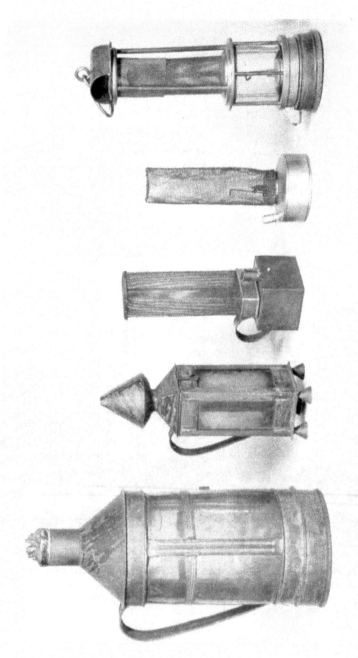

The Miners' Safety Lamp. Some of Davy's own experimental models, with a later modification on the right

overlook the difficulties and embarrassments of life. Luttrell, before I taught him better, imagined muffins grew!' Sir Henry Holland, the famous physician, considered that Lady Holland was acute in distinguishing between real and false merit, and she had a great liking for Jane; she said of her, 'She has more heart, talents, quickness of parts and other excellent qualities than almost anybody. I really *love* her.'

Sydney Smith sometimes used very different terms behind Jane's back from those he used to her face; but he remained her genuine friend, whereas he and Sir Humphry never liked each other. Humphry obviously meant Sydney Smith when he wrote in his notebook of the 'affable jester concealed under the gown of the sacred minister of religion', and Sydney said that Humphry was a 'very foolish coxcomb out of his crucibles'. Crucibles were always in Sydney's eye when he thought of Humphry Davy; he made rather heavy play with crucibles, affinities and decomposition in his early reference to quarrels between Davy and his wife. 'The decomposition of Sir Humphry and Lady Davy is entertaining enough. I wonder what they quarrelled about. He had a very ample allowance of crucible money at marriage. I cannot conceive of any third body interposing to alter their affinities. Perhaps he vaunted above truth the power of chemistry, and persuaded her it had secrets it does not possess – hence her disappointment and fury.' One of the few pleasant remarks Sydney made about Davy was six years later in a letter to Lady Mary Bennet, written from Foston le Clay, where Davy had been one of his many visitors. He wrote: 'I have had a great run of philosophers this summer: – Dr and Mrs Marcet, Sir Humphry Davy and Mr Warburton and divers small mineralogists and chemists. Sir H. D. was really very agreeable – neither witty, eloquent nor sublime, but reasonable and instructive.' Perhaps Humphry tried to be witty like everybody else when he was with the wits, and failed. We know from Lockhart that his conversation was very different when he was with Scott from when he was in London. Scott thought Davy and Jane were equally to blame in the matter of the quarrels: 'She has a temper and Davy has a temper, and these two tempers are not one temper, and they quarrel like cat and dog, which may be good for stirring up the stagnation of domestic life, but they let the world see it, and that is not so well.'

Perhaps it was partly as a solution to domestic difficulties that Davy and his wife decided to renew their continental travels. Davy wrote to his mother in May 1818 that he was going on a very interesting journey and would be absent for some months. The plan was first to visit Flanders and then proceed to Austria and Italy. Evidently Jane told Sydney Smith of their intention of spending some time in the Augsburg Alps, for he wrote to her from Foston, 'I am astonished that a woman of your sense should yield to such an imposture as the Augsburg Alps. Surely you have found out by this time, that God made nothing so curious as human creatures. Deucalion and Pyrrha acted with more wisdom than Sir Humphry and you; for being in the Augsburg Alps, and meeting with a number of specimens they tossed them over their heads, and turned them into men and women. You on the contrary are flinging away your animated beings on Quartz and Felspar.' But off went Jane with Sydney's warning unheeded, and with another flattering letter received just before she started, a letter of farewell. He wrote on May 22nd, 1818:

> My dear Lady Davy,
>   I am *truly* sorry you are going to leave this country, you will be sincerely regretted, and joyfully received back by all good people.
>   As for changing my dear Lady Davy, the only change I shall ever make towards you is to like you better and better . . .

They left England on May 26th, 1818, and did not return till the summer of 1820.

As pleasurable a two years as could possibly be imagined, one would have thought, and we know from *Consolations in Travel* how vividly Davy enjoyed them. The idea, so often expressed, that he would have been better employed following his old beaten track in Albemarle Street seems perverse. Few persons have had a more glorious opportunity to see the beauty of the world and the variety of mankind. Davy used his opportunity to the full, journeying at leisure with eyes and mind alike eager. The last sin to which he was addicted was to be sullen in the sweet air; although he could not, in the nature of things, retain the pure gaiety of his youth, he still had both ready and deep responses to enjoyment.

The principal object of the journey was to visit Naples and try new chemical methods of unrolling the papyri found in the ruins of Herculaneum. But in order to reach their destination the Davys made a wide détour. They visited first the coal-mines in Flanders, where accidents from fire-damp had been so frequent and so severe that gratitude for the benefits of the safety lamp was proportionately strong, and then, travelling sometimes by road and sometimes by riverways, they came to Vienna. The visit to the capital was a concession to Lady Davy's delight in society; the wanderings by river and lake to Davy's pleasure in natural scenery and wide-scale observation. He did not, on this occasion, fish as often as usual; but in passing along the Rhine from Cologne to Coblenz, and in a voyage on the Danube from Ratisbon to Vienna, he was able to make observations for a paper he had in mind, a paper offering facts which might aid in the understanding of the formation of mists in particular situations – especially the formation of mists over the beds of rivers and lakes in calm and clear weather after sunset. Facts had almost become Davy's god; he told his mother at about this time that no letters would please him more than letters from her giving him plenty of facts about the harvests and the fisheries around Penzance; and part of the news he found to tell her was of the good and early harvest in Hungary when he reached that part of the Austrian Empire. He recorded temperatures at stated times of day and night in stated places, giving details of the atmosphere and general weather conditions. His pursuit of facts relative to atmosphere took him to spectacular places at the most dramatic moments, and his record of some of these experiences is most exactly and concisely expressed in his paper communicated to the Royal Society in February 1819, where the scientific purpose restrained his writing.

He did not begin his detailed examination of the papyri at Herculaneum until January 1819. The rolls had been objects of interest for nearly seventy years. The best of the collection had long before been operated on, and those remaining had not only undergone injuries from time but also from other causes, such as transport, rude examination, and mutilation for the purpose of determining if they contained characters. The process which Davy proposed, and which he tried on certain rolls in the museum at Naples, was based on the principle of softening the matter by which the leaves were agglutinated by means of chemical solutions. He was resourceful in experiment and was at first hopeful of success. He wished that Faraday might join him in Naples to assist in the unrolling. But although Davy was provided with government funds to pay an assistant, it was uncertain whether Faraday could return to his work at the Royal Institution if he lightly left it.

He did not go to Italy to help manipulate the papyri; but the closing paragraph of Davy's letter to him, written in October 1818, shows increased informality and esteem. He concluded, 'Mr Hatchett's letter contained praises of you which were very gratifying to me; for, believe me, there is no one more interested in your success and welfare than your sincere well-wisher and friend, H. DAVY.'

Davy's undertaking had been initiated in circumstances of some *réclame*; for the Prince Regent, to whom some papyri had been presented by the Italian Government, was keenly interested in the whole subject, and Davy had gone to Italy as his accredited agent. Dr Thomas Young had for many years been intensely absorbed in problems connected with the unrolling; it was he who had, in Albermarle Street, provided Davy with fragments for his initial experiments. The first results were of an encouraging kind and it was while awaiting a reply to further proposals which Davy sent to the English Government relative to his project that he set out on one of the most enjoyable of all his fishing holidays.

By 19th July he had rejoined Lady Davy at the Baths of Lucca where he analysed the waters. All the world was in Italy, poets and politicians, and 'those who loved to giggle and make giggle'. Byron, by September 1818, had finished the First Canto of *Don Juan* and had dedicated it to Davy's early friend, Southey, 'in good, simple, savage verse, upon the Laureate's politics, and the way he got them'. By February 1819 he was trying to overcome his natural laziness and copy out the Second Canto. He took some trouble to meet Davy in Rome, and lightly saw in him an antithesis to Wellington, master of the 'brain-spattering, wind-pipe-slitting art of war', and to Castlereagh, his hated symbol for those who, in Conspiracy or Congress, had been busy 'cobbling *at manacles* for all mankind'.

Byron saw something hopeful in the era of inventions, though he was ready to have an untrammelled fling at that prospect as at everything else:

> This is the patent-age of new inventions
> For killing bodies, and for saving souls,
> All propagated with the best intentions;
> Sir Humphry Davy's lantern, by which coals
> Are safely mined for in the mode he mentions,
> Timbuctoo travels, voyages to the Poles,

Are ways to benefit mankind, as true,
Perhaps, as shooting them at Waterloo.

Man's a phenomenon, one knows not what,
And wonderful beyond all wondrous measure;
'Tis pity though, in this sublime world, that
Pleasure's a sin, and sometimes sin's a pleasure;
Few mortals know what end they would be at,
But whether glory, power, or love, or treasure,
The path is through perplexing ways, and when
The goal is gained, we die, you know – and then –

What then? – I do not know, no more do you –
And so good night.

Byron had a considerable liking for Lady Davy, but this did not prevent him from making glorious game of all ladies of 'cerulean hue' whenever *Don Juan* offered him opportunity. He liked brows bright with intelligence, and he liked women to have the devil of a spirit; but a show of learning and a reproving piety were his targets. Don Juan's mother [Byron's wife] is described as suffering from both defects; and in a later canto, where Juan goes to London, he describes the 'blues' who descended on him when he made his first appearance in London society. One remembers those 'blue breakfasts' in Blue Square Byron had had with the Davys before leaving London. He writes:

The Blues, that tender tribe, who sigh o'er sonnets,
And with the pages of the last Review
Line the interior of their heads and bonnets,
Advanced in all their azure's highest hue:
They talked bad French or Spanish, and upon its
Late authors asked him for a hint or two;
And which was softest, Russian or Castilian?
And whether in his travels he saw Ilion.

Towards the end of April 1820 Davy went to Ravenna and visited Byron, where he was living in the Palazzo Guiccioli. Both Davy's wife and Byron's mistress had been influenced by Madame de Staël. Lady Davy is reported to have said that before she knew Madame de Staël she was only an ordinary woman; and that to her she owed all her elevation. Byron had made love to Teresa Guiccioli through marked passages in her copy of *Corinne*. Teresa was animated with Davy; but on the occasion of his first visit she made a comic inquiry of Byron as his guest departed. Byron wrote of this to Murray. He told how, after Sir Humphry had gone, 'a pretty young lady of fashion' had turned to him and asked him various questions about his visitor's philosophy. He goes on, 'I explained as well as an oracle his skill in gases, safety lamps, and in ungluing the Pompeian MSS. "But what do you call him?" said she. "A great chemist", quoth I. "What can he do?" repeated the lady. "Almost anything", said I. "Oh, then, *mio caro*, do pray beg him to give me something to dye my eyebrows black. I have tried a thousand things, and the colours all come off; and besides, they don't grow: can't he invent something to make them grow?" ' All this was pronounced, Byron continues, 'with the greatest earnestness; and what you will be surprised at, she is neither ignorant nor a fool, but really well-educated and clever. But they speak like children when first out of their convents; and, after all, this is better than an English blue-stocking.'

Byron adds that he did not tell Sir Humphry this last piece of philosophy, not knowing how he would take it. Lady Davy had remained at Bologna. Byron told Tom Moore that he had intended going off with Sir Humphry to visit her there; but Moore concluded that this intended civility was really designed to give Byron a little breathing space, away from his Contessa. Neither he nor Davy was framed for continued domesticity. In opinion, they differed much. Davy frequently wrote that the grandeur of the Austrian Tyrol gave it superiority over Switzerland; and said that the people were far more agreeable. 'Various in their customs and manners, Illyrians, Italians, or Germans, they have all the same simplicity of character; and are all distinguished by their love of their country, their devotion to their sovereign, the warmth and purity of their faith, their honesty, and (with very few exceptions) I may say their great civility and courtesy to strangers.' Byron had no such illusions about Austrian rule or Austrian intentions. Writing to Murray from Ravenna in September 1820 he said, 'Politics here still savage and uncertain. However we are all in our "bandoliers", to join, the "Highlanders if they cross the Forth", i.e., to crush the Austrians if they cross the Po. The rascals! – and that dog Liverpool, to say their subjects are *happy*! If ever I come back, I'll work some of these ministers.'

The Second Continental Journey

Davy did not share Byron's nationalist enthusiasms; but his own travels, and his liking for good talk, gave him an appreciation of Byron, both as poet and as a spring in the European imagination, denied to his more insular compatriots. When, in 1824, Byron died in the midst of a storm of thunder and lightning, Davy wrote an epitaph not worth quoting; but a poem written a year or so later, with Byron's death still in mind, deserves a place in anthologies of the period. Byron had said, shortly before his death, 'I want to sleep now.' Davy wrote:

> And when the light of life is flying,
>> And darkness round us seems to close,
> Nought do we truly know of dying,
>> Save sinking in a deep repose.

> And as in sweetest, soundest slumber,
>> The mind enjoys its happiest dreams,
> And in the stillest night we number
>> Thousands of worlds in starlight beams;

> So may we hope the undying spirit,
>> In quitting its decaying form,
> Breaks forth new glory to inherit,
>> As lightning from the gloomy storm.

Davy had come into collision with the custodians of the museum at Naples who, at first, had given him and his assistants the facilities required. His paper, '*Some Observations and Experiments on the Papyri found in the Ruins of Herculaneum*', relates the circumstances of the investigation and the progress made. The paper is remarkable for the emphasis laid on the evils befalling ancient recovered treasures from the damp of the atmosphere. Both in writing of the treasures massed in the Louvre, and of the relics preserved in Herculaneum, Davy speaks of the necessity for air-conditioning as a factor in their preservation as though he were a modern curator.

As the scheme for the unrolling of the papyri had not prospered, and Lady Davy was not well, plans were made in the spring of 1820 for a return to England where there was illness also in the Davy family at Penzance. Mrs Davy's sister at Marazion – the Aunt Millett to whom Davy so often sent messages in his letters – had been ill, and Mrs Davy, in visiting her, had suffered from an accident. Davy arranged to return with Lady Davy to England by way of the south of France. He hoped to meet John, who was to be on leave from Ceylon, in London, and to spend with him a holiday at their mother's house at Penzance in the autumn. The plan did not mature. John, after some delay, joined his brother in town, but he went alone for the holiday at Penzance. Humphry was detained in London by two events. He had learnt of the great discovery of the Danish scientist, Oersted, that an electric current could produce magnetic effects; and he had learnt also of the dangerous illness of Sir Joseph Banks. He reached London earlier than he had intended, travelling in advance of Lady Davy, because, he wrote to his mother, of the affairs of the Royal Society. Sir Joseph Banks died on June 20th, 1820. To his mother, four days earlier, Davy had written:

London, June 16, 1820.

My dear Mother – I have been a few days in town, and I am very much disappointed on not finding my brother, who will, I hope, soon arrive. I have a letter from him from the Cape of Good Hope, when he was quite well. Probably they have stopped at St Helena, but I look for him every moment.

I came over in consequence of the illness of Sir Joseph Banks, about the affairs of the Royal Society, rather before the time I intended. Lady Davy is on the road and will join me in a few days.

I have been graciously received by the King.[1] I shall have the pleasure of seeing you before the winter. Pray give my best love to my sisters and aunts, I shall send by the first opportunity some coral necklaces for my sisters, and a shawl for you.

Believe me to be, my dear mother,
Your very affectionate son,

H. DAVY.

He did not go to Penzance in the autumn as he had intended. He wrote to John that he was too busy magnetizing to leave London. When he did go to Penzance it was about a year later, in the late summer of 1821, when he spent ten days in his mother's house, and was honoured by his fellow-townsmen at a public dinner. Apart from his regular correspondence with his family, he had interested himself in such enterprises as the Royal Geological Society in Penzance, founded in 1814. He had subscribed generously to its funds, had

[1] George III had died in June 1820, and the Regent had become George IV.

presented it with specimens he had collected illustrative of the volcanic district of Naples, and had contributed a paper on the geology of Cornwall printed in the first volume of the Society's *Transactions*. Again and again in his lectures and writings one comes upon references to Cornwall, the notices ranging from mineralogy to small-holdings or the colour of trout. The occasion of the public dinner at the Union Hotel in Chapel Street, Penzance, seems to have been an exceptionally happy one. The visit, the renewed intimacy with old friends and relations, recalled Davy not only to the delights but to the thoughts and to the expressions of his youth. He wrote to Poole from Penzance:

Penzance, July 28th, 1821.

My dear Poole,

An uncontrollable necessity has brought me here, close to the Land's End. I am enjoying the majestic in nature, and living over again the days of my infancy and early youth.

The living beings that act upon me are interesting subjects for contemplation. Civilisation has not yet destroyed in their minds the semblance of the Parent of Good.

Nature has done much for the inhabitants of Mounts Bay, by presenting to their senses all things that can awaken in the mind the emotions of greatness and sublimity. She has placed them far from cities, and given them forms of visible and audible beauty.

I am now reviewing old associations, and endeavouring to attach old feeling to a few simple objects.

I am, etc.,

H. DAVY.

15 · *President of the Royal Society*

Davy had been President of the Royal Society for nearly twelve months when he thus looked back at the remoteness and beauty of Mounts Bay, and at what he fancied might be the peace of mind of its inhabitants. His election had not been without incident.

At the time of his death Sir Joseph Banks had been President of the Society for over forty years. The calendar of his manuscript correspondence, recently published, demonstrates the variety of his interests, his staggering industry, and his humorous comprehension of people and affairs, though he was a hard man to cross. He had first been elected President in 1778, the year of Davy's birth; he was re-elected annually until he died. Others besides Dr Thomas Young must have come to consider his cocked hat and fierce eyebrows almost as presidential necessities. He had had his troubles; but he was a character. Gillray proves it by seizing on him and on his sister, Sara Sophia, as subjects for his art. In England it takes a certain quality, which cartoonists unerringly recognize, to rule for years. Sir Joseph was able to deal with people as varied as the man who sent him a seamless coat, and the widow of Lavoisier and Rumford who wanted to have a portrait painted of him in his library. He had thought Davy hardly grave enough to be his successor. 'Davy', he said, 'is a lively and talented man, and a thorough chemist; but . . . he is rather too lively to fill the chair of the Royal Society with that degree of gravity which it is most becoming to assume.'

Davy, though now he was nearly forty-two, must still have seemed very young to Sir Joseph. It was only twenty years earlier that Rumford had written to him that the young lecturer in chemistry at the Royal Institution 'would do well when he had learned not to be idle and not to procrastinate'.

Like most people long in possession of office, Sir Joseph distrusted too much liveliness of mind. He had favoured Wollaston, and Wollaston occupied the chair during the interregnum. The fact that Wollaston, though urged to stand for election by his friends, withdrew his candidature – the Presidency never having been the object of his secret desire – affected Davy during his term of office. He always, in any question as to priority of discovery or suggestion, went out of his way to voice Wollaston's claim and to laud his honour. His letter to Wollaston, preserved in Somerset House, was written when Wollaston had made it unequivocally clear that he would not stand. Davy wrote:

Sunday Evening
June 25th 1820.

My dear Sir,

I solemnly assure you that if on my return from Italy I had found you a candidate for the chair of the Royal Society and supported as you always must be by your friends and by public opinion, nothing would ever have induced me to enter into competition with you.

My feelings were that the only dignity or office which can be the reward of scientific labours ought not to be conferred on wealth and general talent or on mere rank, and it was a sentiment of duty as much as of honest ambition which induced me to offer myself.

I certainly felt severely wounded when I became acquainted with the active and extensive opposition of many of your friends against me at a time when I was almost reposing on the hopes of your support. This wound however you have healed and the manner in which you have done so has increased my admiration of your character.

Should I be elected president I cannot perform the duties of the office or promote the great objects of the Society unless I am supported by the good opinion of yourself and your friends. I hope therefore that no one will retain a feeling of irritation and that if anything foolish or angry has been said on the occasion it will be forgotten.

For myself I can only say that the manner in which you have conducted yourself in the business has added to the very high respect I have always had for your talents, the feelings of gratitude, and greater personal attachment.

I am, My dear Sir
Very sincerely yours
HUMPHRY DAVY.

The Mercurial Chemist

After Wollaston's withdrawal there was little open opposition to Davy; he was elected President on St Andrew's Day, November 30th, 1820.

The best word-portrait we have of him at this time of fulfilled ambition and genial happy mood is in Lockhart's *Life of Scott*. In this same winter of the election to the presidency, Davy was on holiday in Scotland, where he stayed at Abbotsford with, among his fellow-guests, Wollaston and Mackenzie. Lockhart, not an easy judge of men, wrote of him:

> But the most picturesque figure was the illustrious inventor of the Safety Lamp. He had come for his favourite sport of angling . . . and his fisherman's costume – a brown hat with flexible brims, surrounded with line upon line, and innumerable fly-hooks; jack-boots worthy of a Dutch smuggler, and a fustian surtout dabbled with the blood of salmon – made a fine contrast to the smart jackets, white-cord breeches, and well-polished jockey-boots of the less distinguished cavaliers about him. Dr Wollaston was in black, and with his noble serene dignity of countenance might have passed for a sporting archbishop . . . I have seen Sir Humphry in many places, and in company of many different descriptions; but never to such advantage as at Abbotsford. His host and he delighted in each other, and the modesty of their mutual admiration was a memorable spectacle. Davy was by nature a poet – and Scott, though anything but a philosopher in the modern sense of that term, might, I think it very likely have pursued the body of physical science with zeal and success had he fallen in with such an instructor as Sir Humphry would have been to him in early life. Each strove to make the other talk – and they did so in turn more charmingly than I have ever heard either on any other occasion whatsoever. Scott in his romantic narratives touched a deeper chord of feeling than usual, when he had such a listener as Davy; and Davy, when induced to open his views on any question of scientific interest in Scott's presence, did so with a degree of clear energetic eloquence, and with a flow of imagery and illustration, of which neither his habitual tone of table-talk (least of all in London), nor any of his prose writings (except, indeed, the posthumous *Consolations of Travel*) could suggest an adequate notion. I say his prose writings – for who that has read his sublime quatrains on the doctrine of Spinoza can doubt that he

might have united if he had pleased, in some great didactic poem, the vigorous ratiocination of Dryden and the moral majesty of Wordsworth? I remember William Laidlaw whispering to me, one night, when their 'wrapt talk' had kept the circle round the fire until long after the usual bedtime of Abbotsford – 'Gude preserve us! This is a very superior occasion! Eh, sirs!' he added, cocking his eye like a bird, 'I wonder if Shakespeare and Bacon ever met to screw ilk other up.'

Lockhart saw Davy in a more natural light in the country than in London, for Davy was a genuine countryman.

At the time of his election there were those who regretted that the new President should have signally distinguished himself in one particular field of science. 'Sir, we require not an Achilles to fight our battles, but an Agamemnon to command the Greeks', one critic had remarked to Ayrton Paris. That Davy himself came to be partly of this opinion is indicated in a letter to Lady Davy when, in his turn, he came to resign the Presidency. He had determined not to give any opinion to the Fellows of the Society respecting his successor; but he was exceedingly pleased with the idea that Robert Peel, who had shown his zeal to promote science while he was Secretary of State for the Home Department, might be induced to stand. 'He has wealth and influence, and has no scientific glory to awaken jealousy, and may be helpful by his parliamentary talents to men of science.'

Davy came to the Presidency at a time when the country was still suffering from the effects of the war. Sydney Smith in 1820 warned Americans to keep out of war since they would reap not glory but taxation. 'We can inform Jonathan', he wrote, 'what are the inevitable consequences of being too fond of glory; TAXES upon every article which enters into the mouth, or covers the back, or is placed on the foot – taxes upon everything that is pleasant to see, hear, feel, smell, or taste – taxes upon warmth, light and locomotion – taxes on everything on earth, and the waters under the earth – on everything that comes from abroad, or is grown at home – taxes on the raw material – taxes on every fresh value that is added to it by the industry of man.' We know from Davy's notebooks that he was concerned at the political and economic state of the country and oppressed at the confused thinking and gross ignorance too often displayed in public administration and in the services. During his Presidency his general policy was to try to bring about a closer association between government and science, both that scientific research might benefit from the public funds, and that new scientific knowledge might be available to politicians and to the services in making and implementing decisions.

As for the advancement of learning, his vision was one of organized and concerted progress. He had the statesmanlike foresight to see the necessity for organization, without having himself the aptitude of a great administrator. He had been fired quite early in life by Bacon's conception of the College of the Six Days' Works in *New Atlantis* and by the idea of 'trade in light'. He saw the Royal Society, as its founders had conceived it, as a receiving and sorting house, a mart of knowledge, a focus for the learned and inventive of all nationalities, a node for the correlation of all branches of science, so that men, not working in entire isolation and ignorance of what their fellows were doing, might make an organized advance in understanding; and joint progress in utilizing fresh knowledge to the general service of mankind. The application of scientific knowledge could cause destruction and misery; it could also be of untold aid in ameliorating man's lot.

It was with this end in view that Davy tried during his tenure of office to extend the influence of the Royal Society. He knew from experience that private patronage, though of inestimable value, and especially of use in conserving the freedom of an institution from government interference, was insufficient for the large projects he had in mind. He envisaged the establishment of a laboratory for combined research; he wished to bring the Royal Observatory at Greenwich and the British Museum for Natural History into closest association with the Royal Society. As President of the Society he was also a trustee of the British Museum; even during his last illness in Rome he dictated notes (printed in the eighth volume of the *Works*) emphasizing the defects of the Museum as it existed, and drawing up plans for its better future. He dreamed of a Royal Newton College to be founded by George IV. Above all, he wanted persons of diverse interests to mingle.

Towards the close of 1823 he became one of the Founders of the Athenaeum Club. The moving spirit was John Wilson Croker, Secretary to the Admiralty – Byron at one time praised the American Navy to pique him. Croker had proposed a club for literary men and artists. It was Davy who suggested that it should be a scientific as well as a literary club. He replied to Croker's overture:

Nov. 23rd. 1823.

My dear Sir,

We should lose no time in drawing up the 'Prospectus'. I think members of the Royal Antiquarian Society, and of the Linnaean, ought to be admitted by ballot; for my idea is that it should be a scientific as well as a literary club. Lord Aberdeen, with whom I have had a good deal of conversation on the subject, has taken it up warmly.

I know already more than 100 persons who wish to belong to it, and most of their names will be attractive. Mr Heber and Mr Hallum, Mr Colebrooke, Dr Young, Mr Chantrey, Mr Hatchett, Mr Branck, Mr Herschel, and a number of other men of science will give their names.

When I talked to Lord Spencer on the subject, he did not seem to take an interest in it; and Dr Wollaston says he is not a man of clubs. But we are certain of success. The difficulty will be in a short name, and one not liable to any Shandean objections. We can talk of this when I have the pleasure of seeing you tomorrow.

I do not think it would be going too far to make members of the corporate scientific or literary bodies eligible by ballot. I see no reason for excluding Judges, Bishops, and Members of both Houses, none of whom can perform their high duties without a competent knowledge of literature.

Very sincerely yours,

H. DAVY.

Within a month of receiving Davy's letter Croker had named the club, produced a prospectus, and circulated it among the people he wished to bring together. It is said that he objected to the reception of one gentleman as a member on the ground that he was so notorious a bore. Davy's reply is not recorded. The collaboration between the founders was close. Croker was the adept at forming a committee. The maxim 'a great many good names and a few working hands' was his. There is perhaps a hint in the correspondence that he did not altogether trust Davy in the matter of bores. He wrote to him:

December 13th, 1823.

Dear Sir Humphry,

I enclose a few prospectuses of our new Club. I have written the names which I should wish to see of the Committee. In all cases *founders*, as you and I are, must decide who are to be on the Committee; and this is a matter of so great ultimate importance that I would beg of you not to decide on any new name without a consultation. My experience in these matters is considerable, and I assure you that all depends on having a Committee with a great *many* good *names* and a *few* working *hands*.

I am going out of town for one week, but your list of Committee-men may be sent to the Admiralty, and it will reach me in time. I have applied to no one but to Lord Lansdowne, Sir Walter Scott, and Thomas Moore. I wish you would apply to any persons in my list whom you may know; but to avoid mistakes, I shall not apply to anybody else till I hear from you. My list contains about twenty-eight names, the Committee should be of about thirty-six; and we should have four or five practical and practicable people who would attend and help the business. Perhaps a few more artists or a musician should be on the Committee; and what do you say to Charles Kemble? I shall be at home till four.

Yours ever

J. W. C.

The club was opened in temporary quarters in October, 1824; the permanent building in Pall Mall, designed by Decimus Burton, not being completed until after Davy's death. In it his portrait now hangs. According to Croker, 'Irish stews and pancakes' were famous fare in the early days. I suppose Lady Davy never tasted that stew or those pancakes. Davy was able to shut the door safely on himself in that most famous retreat in the world for famous men.

Another project with which Davy was associated during his Presidency of the Royal Society was the founding of the Zoological Society of London. When Sir Stamford Raffles, on his return from the East, considered the possibility of a Society somewhat upon the plan of the Jardin des Plantes, he enlisted Davy's services in his cause. To a cousin he wrote, 'I am much interested at present in establishing a grand Zoological Collection in the Metropolis, with a Society for the introduction of living animals, bearing the same relations to Zoology as a science, that the Horticultural does to Botany. We hope to have 2000 subscribers at £2 each, and it is farther expected, we may go far beyond the Jardin des Plantes at Paris. Sir Humphry Davy and myself are the projectors. And while he looks more to the practical and

Sir Humphry Davy, Bart., President of the Royal Society.
Portrait by John Jackson, R.A., 1823

Dr John Davy. Painted in Malta *c.* 1825 by an unknown Italian artist

immediate utility to the country gentlemen, my attention is more directed to the scientific department.'

Sir Stamford Raffles died in July 1826, not long after the founding of his Society whose existence today bears witness to his vision and to Davy's readiness to help. The names of those first associated with Raffles and Davy in the establishment of the Zoological Society are of interest as illustrating the variety of people with whom Davy was able to exchange ideas. Among others were the names of Davy's friends, the Knights, father and son.

Of inventions which received the initial approval of the Royal Society during Davy's Presidency, one was Charles Babbage's calculating machine. Of this invention Peel wrote to Croker in 1823:

Mr Peel to Mr Croker.

Whitehall, March 8th, 1823.

My dear Croker,

You recollect that a very worthy sea-faring man declared that he had been intimate in his youth with Gulliver, and that he resided (I believe) in the neighbourhood of Blackwall. Davies Gilbert has produced another man who seems to be able to vouch at last for Laputa. Gilbert proposes that I should refer the enclosed to the Council of the Royal Society, with a view of their making such a report as shall induce the House of Commons to construct at the public charge a scientific automaton, which, if it can calculate what Mr Babbage says it can, may be employed to the destruction of Hume. I presume you must at the Admiralty have heard of this proposal.

'Aut haec in vestros fabricata est machina muros,
Aut aliquis latet error.'

I should like a little previous consideration before I move in a thin house of country gentlemen, a large vote for the creation of a wooden man to calculate tables from the formula $x^2 + x + 41$. I fancy Lethbridge's face on being called on to contribute,

Ever affectionately

ROB PEEL.

To this letter Croker replied:

March 21st, 1823.

My dear Peel,

Mr Babbage's invention is at first sight incredible, but if you will

o [193]

recollect those little numeral locks which one has seen in France, in which a series of numbers are written on a succession of wheels, you will have some idea of the first principles of this machine, which is very ingenious and curious and which not only will calculate all regular series, but also arranges the types for printing all the figures. At present indeed it is a matter more of curiosity than use, and I believe some good judges doubt whether it ever can be of any. But when I consider what were called Napier's bones and Gunter's scale, and the infinite and undiscovered variety of what may be called the *mechanical powers* of numbers, I cannot but admit the possibility, nay the probability, that important consequences may be ultimately derived from Mr Babbage's principle. As to Mr Gilbert's proposition of having a new machine constructed, I am rather inclined (with deference to his very superior judgement in such matters) to doubt whether that would be the most useful application of public money towards this object at present . . .

Croker counselled Peel to refer the matter to the Council of the Royal Society. He thought the scientific members would give the best opinion of the value of the invention and, that obtained, it could be considered whether another machine should be made at the public expense, or whether Mr Babbage should receive a reward either from Parliament or the Board of Longitude.

On April 1st, 1823, the scheme was officially submitted to the judgment of the Royal Society. It was favourably reported on, and Babbage was aided by a grant of £1,500 from the Civil Contingencies Fund. After four years Babbage went to Italy where much of his later life was spent. The subsequent history of his invention is most curious; it belongs to the story of Ada Byron, daughter of Byron's 'princess of parallelograms'. She, like her mother, was a mathematician.

Babbage was a difficult character and had no great cause to love Davy who preferred J. G. Children before him in the matter of a secretaryship. In his book, *The Decline of Science in England*, published in 1830, he went so far as to say of Davy that he hoped the writings of the philosopher might enable his contemporaries to forget some of the deeds of the President of the Royal Society. His analysis of the characters of Wollaston and Davy is the more valuable as made by one who was far more apt to cavil at Davy than to flatter his memory. After noting the extreme caution of Wollaston he wrote of Davy:

> Ambition constituted a far greater ingredient in the character of Davy, and with the daring hand of genius he grasped even the remotest conclusions to which a theory led him. He seemed to think invention a more common attribute than it really is, and hastened, as soon as he was in possession of a new fact or principle to communicate it to the world, doubtful perhaps lest he might be anticipated; but confident in his own powers, he was content to give to others a chance of reaping some part of that harvest, the largest portion of which he knew must fall to his own share.
>
> Dr Wollaston, on the other hand, appreciated more truly the rarity of the inventive faculty; and, undeterred by the fear of being anticipated, when he had contrived a new instrument, or detected a new principle, he brought all the information that he could collect from others, or which arose from his own reflection, to bear upon it for years, before he delivered it to the world.
>
> The most singular characteristic of Wollaston's mind was the plain distinct line which separated what he knew from what he did not know; and this again, arising from his precision, might be traced to caution . . .
>
> In associating with Wollaston you perceived that the predominant principle was to avoid error; in the society of Davy, you saw that it was the desire to see and make known truth. Wollaston never could have been a poet; Davy might have been a great one.

Davy's strong desire was for co-ordination of effort. Just as the wise governors of New Atlantis sent learned men to search out and bring back knowledge from all parts of the world, so Davy, like Sir Joseph Banks, wished the Society to promote expeditions; he supported the Polar expedition of his day. He welcomed returned travellers. He was not careless of the arts. The sculptor Chantrey loved to talk with Davy and Wollaston.

But – 'Give me that patience, patience that I lack', might well have been Davy's cry. Patience is, I suppose, above all qualities necessary for an administrator, and Davy, patient with nature, and capable of nursing and testing an idea through a long period of years, was impatient of man. He had an outward-looking rather than an inward-looking eye. He did not know himself or others through himself. The imputation of vanity cannot, I think, be sustained by anyone who carefully reads the records of this period of his life; but he came to be greatly harassed and to suffer vexation of heart. His very lack of sustained vanity made things hard for him since, as Franklin said, a man should thank God for his vanity among the other comforts of life. Davy was excessively vulnerable, open, incapable of finesse. He was not averse to dressing up and was, indeed, sufficiently theatrical – every good lecturer or preacher must have some of the qualities of an actor. But the part he was least fitted to play was the part of 'a ful solempne man'. He could not keep the mask adjusted. Even his speeches before the Royal Society have not the same ring as the discourses which prefaced his lecture-series at the Royal Institution; they are not less earnest, but they are further from the immediate reality of his own feeling, more paraphrastical and inclining to the pomp of peroration.

Davy's tendency to be edifying was not always kept in check. His Address to the Society delivered after his election remains inspiriting. But in a letter signed J. W. L. G., printed as a pamphlet in 1821, he was told, 'Leave your Royal Institution style (however excellent of its kind) as an heir-loom to Mr Brande.' The writer then continued, 'You fell into the error I had supposed you would, and though I think you did not say one whit too much of yourself, yet there was a glare and ornament in the style, and a fastidiousness in the manner which, however it might have suited the audience in the lecture room in Albemarle Street, was objectionable, in the highest degree, in the President of the Royal Society.'

Flowers amidst the corn had been anathema to the Royal Society from its inception. To Davy, who had a strong picture-making faculty, and who thought in analogies, the prospect was lenten. The whole pamphlet is in a censorious tone, though the new President had hardly assumed office. Had Davy known a line written by Anna Beddoes' son he might often have said,

> But there's a blasting thought stirring among you.

In subsequent Anniversary Addresses, delivered on successive St Andrew's Days, on the occasion of presenting the medals upon Sir Godfrey Copley's donation, Davy had to sum up the merits of those honoured, and bestow the approbation of the Royal Society.

There was always a danger that in praising people to their faces, although he was praising them as the representative of the Royal

Society and not in his own person as a fellow-scientist, he might assume a tone which sounded patronizing and condescending. This was a particular problem when the first Royal Medal was awarded. In 1825 at the anniversary dinner of the Royal Society, Robert Peel, at that time Secretary of State for the Home Department, announced King George IV's intention of founding two annual prizes, each of the value of fifty guineas, to be at the disposal of the President and Council of the Royal Society. The prizes were established in the form of silver and gold medals, to be given for important discoveries or useful labours in any department of science; and they were open to learned and ingenious men of all countries. Davy and the Council awarded the first prize, in the first year (1826), to 'Mr John Dalton, of Manchester, Fellow of the Royal Society, for the development of the Chemical Theory of Definite Proportions, usually called the Atomic Theory, and for his various other labours and discoveries in physical and chemical science'.

Davy was in a difficult position. He had been converted to the theory of definite proportions, but not to the doctrine of indivisible atoms. To the end of his days he spoke of 'corpuscles' or 'particles' of 'bodies *believed* to be elementary'. At about the time when he first visited Rome he wrote in a notebook, 'Men generally find it most easy to explain everything who are ignorant of everything. The progress of physical science is slow but not sure. Probably the atomic philosophers had an idea exactly like that of the moderns; Democritus, the pure atomic doctrine; Pythagoras, the doctrine of definite proportions and *regular* forms.'

Before presenting the medal to Dalton, Davy gave a brief sketch both of prior conceptions and of loose notions on this same mode of viewing the combinations of bodies; but he showed that statistical chemistry as it was taught in 1799 was 'obscure, vague, and indefinite, not meriting even the name of a science', and that to Mr Dalton belonged the distinction of first unequivocally calling the attention of philosophers to these important subjects. He then reverted to prior conceptions and ingenious illustrations and continued:

Mr Dalton's permanent reputation will rest upon his having discovered a simple principle, universally applicable to the facts of chemistry – in fixing the proportions in which bodies combine, and thus laying the foundation for future labours, respecting

[197]

the sublime and transcendental parts of the science of corpuscular motion. His merits, in this respect, resemble those of Kepler in astronomy. The causes of chemical change are as yet unknown, and the laws by which they are governed; but in their connection with electrical and magnetic phenomena, there is a gleam of light pointing to a new dawn in science; and may we not hope that, in another century, chemistry having, as it were, *passed under the dominion of the mathematical sciences*, may find some happy genius, similar in intellectual powers to the highest and immortal ornament of this Society, capable of unfolding its wonderful and mysterious forms.

The King, it is said, can do no wrong; the lay-reader of the history of the Royal Society comes to the conclusion that the President can do no right. In Sir Joseph Banks' day there were factions; in Davy's day there were factions; in the day of his successor, Davies Gilbert, there was to be open war. Charles Babbage said that the decision of the Society to award the first Royal Medal to Dalton was an insult to that distinguished man.

Members of the general public were anxious to read the Discourses. Davy had remained 'kind and constant' in his friendship with the Edgeworths. Maria Edgeworth, before the Discourses were published, had the courage, or the rashness, as she said, to ask for a particular manuscript to read. Davy lent her not one but all of them. She said they fully came up to her expectations. To her they gave a complete account of the progress and discoveries of science during the six years. She thought they had the dignity of perfect simplicity and candour, and that Davy expressed himself as one sensible of the national glory but free from national jealousy, since his great object as a philosopher was the general advancement of science. She thought he enjoyed giving praise where praise was due. His addresses to the medallists she considered noble; 'always appreciating the past with generous satisfaction, yet continually exciting to future exertion'.

In 1812, when Sir Joseph Banks had heard of Davy's intended marriage, he had laughingly said that it only needed the countenance of the ladies to make science really fashionable. But perhaps Davy was nervous of his lady's sharp tongue in company. Describing a party at his house in December 1820, Henry Edward Fox wrote, 'Went with Mrs Ord to Lady Davy's, where we found Lady Holland in state upon the sofa, and the ugly Abercromby administering her spawny flattery in more than usual studied phraseology . . . Lady Davy was so anxious and fidgety that she could hardly sit still or find time to scold Sir Humphry.' In 1822 Maria Edgeworth described him as a martyr to matrimony. Although Davy followed, at first, the example of his predecessor in inviting to his house, during the session, members of the Royal Society and kindred guests, these evening parties remained exclusively masculine. They were held on Saturday instead of as formerly on Sunday evenings. John Davy says that while he was in England they brought together 'not only men of science, but also literary men, poets, artists, country gentlemen, and men of rank, and they were very attractive to foreigners. The subjects of interest of the day were discussed, curious information obtained from the best sources, and knowledge exchanged between individuals.' But there were troubles. One can well imagine a man as careless as Davy being casual about the invitations, the consequence being that some people, far from being gratified, nursed wounded susceptibilities. They said the President had too high an opinion of himself and had learnt to be above his connexions.

When the Davys moved from Grosvenor Street to a sunnier and better-placed house, 26 Park Street, Grosvenor Square, the evening parties, from which Davy had hoped so much in cementing diverse interests among the learned, were discontinued. Instead, the library of the Royal Society in its apartments at Somerset House was opened on Thursday evenings, after the regular meeting was concluded, where the Fellows and visitors could converse familiarly on matters of science.

As for Lady Davy, she had plenty of compensations. She found literary and political lions more amusing than their scientific counterparts, and she had never needed the prestige of being the wife of the President of the Royal Society to give her social consequence. Sydney Smith, and the Earl of Dudley, were considerably more to her taste. Of Dudley, who must have been a most amusing companion, she became the firm friend. One of her best means of entertaining others in later life was by recounting stories of his absent-mindedness. There could have been no greater contrast than between Davy, who could make up his mind and act on it in an instant, and Dudley, who in any matter of choice, even in the choice between two words, remained for long agonizingly balanced. It was he who is said to have consulted his colleagues as to whether, in a diplomatic sentence, he should perhaps insert the word 'perhaps'. But then, Dudley had a humorous eye to his own foibles, a gift that had not been bestowed on Davy. And he had sound instinct. There was no 'perhaps' impeding his judgment when he exclaimed, on hearing in 1826 that Lady Davy's cousin was ruined, 'Scott ruined! The author of Waverley ruined! Good God, let every man to whom he has given months of delight give him a sixpence, and he will rise tomorrow morning richer than Rothschild!'

Some of Davy's troubles as President of the Royal Society were inherent in the situation. In an orchestra of musicians it is physically impossible for the conductor to play first violin. But the choice allowed to any Society in choosing a President must often be between some amateur of rank or influence and its own first fiddle. Conductors of genius who affect the very tone of the instruments are rare birds.

Davy's papers read before the Royal Society in the years 1820–6 were printed in the *Philosophical Transactions* and are reprinted in the sixth volume of his collected *Works*. Beginning with the letter to Dr Wollaston, *On the Magnetic Phenomena Produced by Electricity*, read before the Society on November 16th, 1820, and concluding with his last Bakerian Lecture, *On the Relation of Electrical and Chemical Changes*, read on June 8th, 1826, they demonstrate the continuity of his researches, and how this continuity was extended by the grafting of Faraday's mind to his master's.

Just as Davy had stated the decomposition of water by Nicholson and Carlisle in 1800 to be the cardinal fact leading to his earlier discoveries, so he said Hans Christian Oersted's discovery of the true connexion between electricity and magnetism, revealed in the winter of 1819, was cardinal to the electrochemical researches he undertook between 1820 and 1826. Scientists were hot on a fresh trail.

Wollaston was the first to deduce the possibility of electromagnetic rotation. He perceived that there was a power not directed to or from the wire in the voltaic circuit, but acting circumferentially round its axis. He wagered that he would make the wire revolve on itself; the wager was open. Faraday knew of it as well as other interested persons. In April 1821 Wollaston, with Davy helping him, tried experimentally to produce this rotating motion, but failed. Faraday did not see the apparatus or the experiment; but, coming in later, he did hear the conversation of Wollaston and Davy, and he knew Wollaston's expectation that he would make a wire in the voltaic circuit revolve on its own axis.

During the summer of that year Davy, as was always his custom, went out of town. Wollaston, too, was out of town. Faraday was engaged in writing an historical sketch of electromagnetism. By repeating all the published experiments and initiating others himself, he was the first to discover, in the beginning of September 1821, the rotation of a wire in the voltaic circuit round a magnet, and of a magnet round the wire. He published his result before the end of September without reference to Wollaston. In October, Davy and Wollaston returned to London.

Davy was President of the Royal Society; Warburton and other friends of Wollaston were certainly incensed against his protégé. In subsequent publications various statements were made in which Wollaston's expectation was noticed, and in 1823, upon Faraday's proposed election to the Royal Society, an historical statement was made by him with reference to Wollaston in the Royal Institution *Journal*. There the matter might have rested; Faraday was merely over-hasty. One's sympathies are with him. At the time it was he who suffered; but historically Davy has been more the injured one, because it is constantly assumed that he alone took the lead in opposing Faraday's election to the Royal Society in 1823; whereas Warburton wrote to Faraday plainly in a letter that he had told the President that, while not seeking to oppose Faraday's election on the score of intellectual achievement, he should take the opportunity to say in public what he thought of his (Faraday's) action in relation to Wollaston. It was not until he had read Faraday's historic statement that he wrote to him in July, 1823:

18 Lower Cadogan Place,
July 8, 1823.

Sir, I have read the article in the Royal Institution Journal (vol. xv. p. 288) on electro-magnetic rotation; and without meaning to convey to you that I approve of it unreservedly, I beg to say that, upon the whole, it satisfies me, as I think it will Dr Wollaston's other friends.

Having everywhere admitted and maintained that on the score of scientific merit you were entitled to a place in the Royal Society, I never cared to prevent your election, nor should I have taken any pains to form a party in private to oppose you. What I should have done would have been to take the opportunity which the proposing to ballot for you would have afforded me to make remarks in public on that part of your conduct to which I objected. Of this I made no secret, having intimated my intention to some of those from whom I knew you would hear of it, and to the President himself.

Further Researches: A Northern Tour

When I meet with any of those in whose presence such conversation may have passed, I shall state that my objections to you as Fellow are and ought to be withdrawn, and that I now wish to forward your election.

I am, Sir, your faithful servant

HENRY WARBURTON.

Wollaston too accepted Faraday's justification of himself, but Davy continued to oppose his election to the Royal Society. He was elected on January 8th, 1824. Faraday was very sensitive to opinion; but not very imaginative about other people. Davy on no occasion defended himself against aspersions though he showed his anger at them. That he had cherished the idea of training young men to succeed or surpass him is apparent in his attitude towards Edmund and John Davy and, in another direction, towards the son of his old friend – Andrew Knight. In a poem in which he described two eagles teaching young eagles to fly, an incident which he witnessed in a remote part of the Highlands in August 1821, he wrote of his hopes:

> The mighty birds still upward rose
> In slow but constant and most steady flight,
> The young ones following; and they would pause,
> As if to teach them how to bear the light,
> And keep the solar glory full in sight.
> So went they on till, from excessive pain,
> I could no longer bear the scorching rays;
> And when I looked again they were not seen,
> Lost in the brightness of the solar blaze.
> Their memory left a type and a desire:
> So should I wish towards the light to rise,
> Instructing younger spirits to aspire
> Where I could never reach amidst the skies,
> And joy below to see them lifted higher,
> Seeking the light of purest glory's prize:
> So would I look on splendour's brightest day
> With an undazzled eye, and steadily
> Soar upward full in the immortal ray
> Through the blue depths of the unbounded sky,
> Portraying wisdom's matchless purity;

The Mercurial Chemist

Before me still a lingering ray appears,
But broken and prismatic seen through tears,
The light of joy and immortality.

Aspiration such as this was in the air he breathed. Keats in 1819 built his epic fragment of *Hyperion* on the idea that the new gods, by exceeding the old in glory, rightly vanquished them. Davy had the magnanimity to see that knowledge, and the power that comes of it, should be given away freely. In writing an obituary of Playfair, after praising his intellectual power, the absence in him of all prejudice, and his language 'open and candid as his soul', Davy emphasized that 'malignity, envy, and uneasiness at the success or talents of others never tainted his character'. It is possible to praise virtues without possessing them, and to act on impulses conceived in the obscure and jealous depths of the being. But Davy had been consistently helpful to Faraday. It was he who had encouraged the younger man to think he might do something independently for science; he who had corrected his solecisms; he who had spoken to Earl Spencer to obtain for him married quarters at the Royal Institution when he wished to marry.

When the pupil becomes independent of the master, there must almost of necessity be a painful operation. From being 'directed' or 'requested' to perform experiments and report on them, Faraday gradually worked free. He might have saved himself some pangs if he had delayed a little to declare, as it were, his majority. At about the same period as he was working at electromagnetism he was working also on chlorine. Since 1813 he had been, in Davy's varied researches, his operator and assistant, and his aid had been acknowledged by Davy repeatedly in published communications, not with the warmth with which he had acknowledged his brother John's help on one or two occasions, but with punctilious care. Davy was still honorary professor of Chemistry at the Royal Institution, Brande being professor of Chemistry and Faraday chemical assistant. The story of the liquefaction of chlorine gas by Faraday was told with insinuations against Davy by Ayrton Paris who witnessed the original experiment. He hinted that Davy filched his pupil's subject.[1]

---

[1] The effect of Paris's insinuations was to put John Davy on the defensive. He spoilt his earliest biography of his brother with refutations. Controversy has continued.

Further Researches: A Northern Tour

Faraday had been working on chlorine; but it was Davy who had suggested the particular experiment which Paris witnessed, and it was never Davy's custom to state explicitly the result he anticipated from an experiment. Faraday says later that a fellow-experimenter is only of use if he does not know what to expect. Otherwise his view of the result is likely to be prejudiced in favour. Faraday had produced nothing of special value in the paper he had already written on his experiments with chlorine until Davy, with his wide knowledge, played his customary part of prompter. Faraday's industry and ingenuity were used in further experiments on the liquefaction of gases under Davy's general direction, and the result was the paper 'On the Application of Liquids formed by the condensation of gases as mechanical agents' read before the Royal Society on April 17th, 1823.

Davy opened his paper with:

One of the principal objects I had in view, in causing experiments to be made on the condensation of different gaseous bodies, by generating them under pressure was the hope of obtaining vapours, which, from the facility with which their elastic forces might be diminished or increased by small decrements or increments of temperature, would be applicable to the same purposes as steam.

Towards the close of the paper he wrote:

I hope soon to be able to repeat these experiments, in a more minute and accurate way; but the general results appear so worthy of the attention of practical mechanics, that I think it a duty to lose no time in bringing them forward in their present imperfect state.

John Davy's footnote on the author's intention to repeat the experiments in a more minute and accurate way, as often with his comments on his brother's aims, is illuminating. He wrote (in 1840):

This intention, I believe, the author did not carry into effect, nor has it been attempted that I am aware of by others, notwithstanding the incalculable importance of the object in view – the obtaining of a fluid applicable to the steam engine, capable of producing the effects of ordinary steam, without the consumption of fuel. He was diverted from this inquiry by another very important one, which about this time he engaged in, namely, the protection of the copper sheathing of ships.

During the War of American Independence copper sheathing had been applied for the first time to ships' bottoms to protect the wood from destruction by worms and to prevent the adhesion of weeds, barnacles, and other shell-fish. The rapid decay of this copper sheathing and the uncertainty of the length of its duration had for a long time occupied the minds of those most concerned in the naval interests of the country. In 1823 Davy's attention was called to the subject by the Commissioners of the Navy Board. A Committee of the Royal Society was chosen to consider it, and Davy entered into an experimental investigation. His papers on the subject show how he proceeded from hypothesis, through experiment, to application on a small scale, and application on an extended scale, back again to general principles, seeking always to illustrate obscure parts of electrochemical science by the ascertaining of new facts.

In his first paper, read before the Royal Society, on January 22nd, 1824, on the corrosion of copper sheathing by sea-water, Davy describes experiments on specimens of copper sent by John Vivian to Faraday and on specimens collected by the Navy Board for the Royal Society. He concluded from these experiments that copper might be preserved from corrosion by sea-water by being rendered electrically negative, and that this might be effected by a simple method – by bringing copper artificially into contact with a more oxidizable metal. After many experiments he suggested that small quantities of zinc, or, much cheaper, of malleable or cast iron, placed in contact with the copper sheathing of ships, all in electrical connexion, would entirely prevent corrosion. The ocean might be considered, in its relation to the quantity of copper in a ship, as an infinitely extended conductor. He supposed, too, that as negative electricity could not be supposed favourable to animal or vegetable life, and as it occasioned the deposition of magnesia, a substance noxious to land vegetables, upon the copper surface, and as it must assist in preserving its polish, that there was considerable ground for hoping that the same application would keep the bottom of ships clean, a circumstance of great importance both in trade and in naval war.

During the next six months Davy was given facilities by Lord Melville and the Lords of the Admiralty, through the Commissioners of the Navy Board and the Dock Yards, to conduct experiments upon a very extended scale at Chatham and Portsmouth. In a very short notice of these experiments read before the Royal Society on June 17th, 1824, it seemed to Davy that his results were satisfactory and conclusive, and even surpassed his expectations. Cast-iron protectors, he found, lasted longer; and cast iron was cheaper than malleable iron or zinc. He experimented on the limits of protection, and found the application of a very small quantity of the oxidizable metal more advantageous than a larger one. During the course of the work many singular facts connected with general science were briefly noted by him, and the possible application of those facts to the preservation of animal and vegetable substances. In concluding his paper he stated how much he owed to the care and attention and accuracy of Mr Nolloth, Master Shipwright, and Mr Goodrich, Mechanist in the Dockyard at Portsmouth, in superintending the execution of many of the experiments.

In the paper read to the Society a year later, June 9th, 1825, the focus of interest shifts from the preserving of the copper sheeting to the prevention of fouling. That copper was preserved by the protectors had been proved; but the no-less-important circumstance – that this preservation should keep the ships' bottoms free from adhesions of weed and shell-fish – was in doubt. In his last Bakerian Lecture on June 8th, 1826, in an historical survey 'On the Relations of Electrical and Chemical Changes', Davy wrote:

A great variety of experiments made in different parts of the world has proved the full efficiency of the electro-chemical means of preserving metals, particularly the copper sheathing of ships; but a hope I had once indulged, that the peculiar electrical state would prevent the adhesion of weeds or insects, has not been realised; protected ships have often, indeed, returned after long voyages perfectly bright, and cleaner than unprotected ships, yet this is not always the case: and an absolute remedy for adhesions is to be sought for by other more refined means of protection, and which appear to be indicated by these researches.

It must have been bitter for him to make this admission. We know from his letters to John and to his mother how near his heart was the project, how generous his hopes, and how bright his anger at the gloating, premonitory newspaper reports of failure, and the gibes which went with them – gibes of foolish confident men. At the outset he had written to John, 'I have found a simple method of preserving the copper-sheeting of ships, which now rapidly corrodes. It is by rendering it negatively electrical. The results are of the most beautiful and unequivocal kind . . . I was led to this discovery by principle, as you will easily imagine; and the saving to government and the country by it will be immense. I am going to apply it immediately to the navy. I might have made an immense fortune by a patent for this discovery; but I have given it to my country, for in everything connected with interest, I am resolved to live and die at least *sans tache*.' When premature reports of failure appeared in the Portsmouth papers he wrote to his mother, 'Do not mind any lies you may see in the newspapers, copied from a Portsmouth paper, about the failure of one of my experiments. All the experiments are successful, more even than I could have hoped.'

Letters which passed between J. G. Children and Sir John (then Mr) Barrow of the Admiralty give what Children's biographer calls 'an amusing comment on the wanton disregard for truth and absence of good feeling too often disgracing newspaper assertion'. Children's letter was dated October 22nd, 1824, and was addressed to Mr Barrow from the British Museum:[1]

My dear Sir,
   You have seen, no doubt, a paragraph in The Times newspaper of the 16th instant, stating that vessels coppered on Sir Humphry Davy's plan with protectors, have returned after short voyages perfectly foul! In the same paragraph it is also insinuated that Sir H. D's late voyage to the Baltic was made at the public expense. Pray allow me to ask you if these statements, or either of them, be correct or otherwise?

I am, my dear Sir,
Your faithful servant
J. G. CHILDREN.

To this letter Mr Barrow replied:

Admiralty, October 22nd, 1824.

My dear Sir,
   In answer to your inquiries respecting vessels coppered on Sir Humphry Davy's plan, with protectors, having returned from short

---

[1] *Memoir of J. G. Children*, Anna Atkins, 1853, p. 228.

voyages perfectly foul, and whether Sir Humphry Davy made a voyage to the Baltic at the public expense, I have to state that with regard to the former no report whatever has been received at this office from any one vessel supplied with protectors, nor am I aware that any one of them has returned into port. And with regard to the second point, I can safely say that Sir Humphry's passage to the Naze of Norway (not the Baltic) was not attended with any expense to this or any other department of government. The fact is simply this: the Comet steam-vessel having been ordered to proceed to Heligoland at the express request of the King of Denmark, for the purpose of fixing with precision by means of numerous chronometers, the longitude of that island, in order to connect the Danish with the British survey, and the Board of Longitude having recommended that the voyage should be extended as far as the Naze of Norway, for the purpose of ascertaining the Longitude of that important point, Sir Humphry Davy volunteered to proceed in her, at his own expense, to enable him to attend in person to certain experiments which he was desirous of making on the action of sea-water on the copper of a vessel passing rapidly through that medium.

If any illiberal construction should have been conveyed to the public, as your note would seem to imply, you are at liberty to make use of this reply in any way you may deem fit.

I am, my dear Sir,

Very sincerely yours,

JOHN BARROW.

The expedition commented on in this letter, and which Davy undertook voluntarily in one of the earliest survey steamships, was in charge of the astronomer, Dr John Lewis Tiarks (1789–1837). Lord Clifton accompanied Davy, who also made of the voyage an opportunity for seeing something of the Scandinavian countries. He wrote an account of this seven weeks' tour while he was ill at Gothenburg. Details also appear in his letter to his mother and to Lady Davy who was with a party in Switzerland. Women with any refinement could hardly, Davy thought, travel in the North.

The party left London on June 30th, 1824. Davy saw for the first time the whole of London River, and marvelled at the 'immensity of British capital industry, and activity displayed by the great inlet to the most wonderful city in the world'. The weather was fair to Heligoland; but on their landing the rain fell in torrents. Davy found life miserable on this rude island. The houses in general were low and confined, but all articles of life were very cheap, particularly bad wine and spirits. Madeira and brandy were a shilling a bottle. During the journey from Heligoland to Norway the weather was stormy. 'Towards night the ship rolled considerably; the water dashed over the decks, the vaunted power of steam over the elements was as nothing: yet the steam carried us on, though slowly; and after much labouring and some danger, and much dripping, and one death, that of an unfortunate painter, we made the coast of Norway at seven in the morning – my experiments ruined, a misfortune to add to the misery of sickness.'

At Kristiansand he was wonderfully entertained. The story is much of food. The most amusingly described dinner is with Mr Mark, the English Vice-Consul, who took his guests to a country house he owned up the river, promising that Davy should catch a salmon. Davy was, as usual, absorbed by the sight of the waterfall – 'one of the great machines of nature' – and ready to enjoy the dinner. The prelude to this feast was kippered salmon, anchovies, brown bread and butter, and various liqueurs; the first course was ham, and peas boiled with sugar in their shells; then some salmon boiled; then chickens roasted with plenty of parsley in their bellies; then roast veal; and last of all cranberry jelly, with cakes and sweet things. There was fruit on the table which seemed as a garnish; and salad with cheese, the salad being especially good. At Arendal, where Davy went to visit the iron-mines, he went to Mr Tiddicamp's country seat for a feast – a dinner to which all the neighbourhood was invited. There were toasts during the whole dinner. When Davy gave '*Liberty*, FREYHEIT', the whole party rose and sang a song in full chorus. Davy's health was drunk. The Royal Society was toasted, together with the British Constitution, and the memory of Lord Byron. After dinner they all shook hands, and then walked to see a most magnificent view; the sea on one side, and wood almost interminable, with lake and mountains, on the other and a thousand little ponds all surrounded with woods.

At Arendal the English visitors were struck by the manner in which the women worked. Davy says the postmaster was rowed to the Vice-Consul's, to the grand dinner, by a female servant who was rather good-looking and young, and who dashed through the surge as a Thames boatman would have done, with her great hulking master sitting opposite her. Davy himself was carried across the lake from the iron-mines by a boat-woman.

At one dinner in Norway, when Davy's host had proposed the toast, 'The British Constitution as a model for all the world', Davy had responded by proposing, 'Norwegian hospitality as a model for all the world'. He would not have proposed a similar toast to Swedish hospitality. The first Swedish inn at Strömstad was a bad sample of an indifferent succession.

At Gothenburg he was received by the Crown Prince of Sweden, and dined with him and the Princess, granddaughter of the Empress Josephine. He thought the upper countenance of the Princess beautiful; he admired her fine blue eyes and perfect complexion; and found her graceful and gracious in person and manner; but in constitution feeble. The Prince, with whom he talked much after dinner, on a great variety of subjects, seemed to have enlightened views on education and the policy of kings; he understood something of chemistry and had general views of all the sciences. In later life when the Crown Prince wrote Davy an affectionate letter Davy confessed to Lady Davy 'that it was in human nature to be pleased with the recollection of the heir to a throne'. But while in Sweden his mind was far more engaged with the double snipe.

From Gothenburg he drove to Falkenberg and Hälsingborg, where he had an interview with Berzelius, whom he found in good plight, rather fatter than when he had seen him twelve years earlier in London. He then hired a boat and in four hours landed in the capital of Denmark. Here he met Professor Oersted, who showed him his apparatus for increasing thermo-electro-magnetism. By Prince Christian of Denmark he was received in the kindest manner, without ceremony, and then entertained at dinner, Oersted being of the party. He was given leave by the Prince to shoot at Saltholm, where it was said double as well as single and jack snipe bred. He found an immense variety of wading birds on this low and flat island. On Sunday he dined again with the Prince; but this was a visit of ceremony; – 'thirty people, and of course rather a bore. Went away as soon as possible to fish.' On Tuesday, he took the Kiel boat, and had a pleasant sail to Kiel through the islands, and a most wretched journey from Kiel to Hamburg through sands and deserts. The Danes and Holsteiners appeared to him to be rather fat-headed – 'a feeding and smoking people'. The excursion which he made to Altona and Bremen gave him most pleasure. In the course of his visit he met Schumacher, Gauss, and Olbers, who showed him the telescope with which he discovered his two new planets. Of these meetings he wrote, 'I am rejoiced that I made the excursion to Altona and Bremen: it has given me a better idea of human nature; for Schumacher, Olbers and Gauss appear to me no less amiable as men than distinguished as philosophers; they have all the simplicity, goodness of heart and urbanity of manners which ought to make us proud of the name, and of the influence of intellect and scientific pursuits upon the morals, and habits, and the affections.' With these three celebrated men he said he spent one of the most agreeable days belonging to the later period of his life.

He returned to England in the steamship, again experiencing stormy weather and sickness. He always said the sea was a glorious dominion but a wretched habitation. Had a paddle given way in the storm, he told Lady Davy, nothing could have saved them. Throughout his tour he had been honoured and fêted and given every opportunity for sport, and for the study of natural history in which he delighted. Not all his experiments had suffered from the caprices of the sea and wind. Yet one notices in all his accounts of this journey a decline in spirits not due merely to physical exhaustion and frequent sickness. His verse at this time turns from enthusiastic delight in all created things to the consideration of mortality.

When he saw in Germany how the grave of Klopstock whom, like many of his contemporaries, he considered a great poet, was forsaken and unknown, he lamented in a letter to Lady Davy that fame did not last; it was merely the passing glory of an hour. When he wrote this letter he and Jane were separated by the whole length of the Elbe. She was merry-making near its source and he was ruminating near the mouth. She might, he said, have floated a letter down to him.

17 · *The Beginning of Illness:* Salmonia

It says something for Davy's prose, and for his way of living in the North, that he managed to convey the character and extent of the countries he visited, the sound of birds, the wind and the waste. He wrote to Poole that the North, if not a kind, was a beautiful mother; he wondered that she had not produced more poets. He had travelled by sea and land, he told his own mother, more than two thousand miles. Both Mrs Davy's sons were still seeing the world. While Humphry was in the North, John was in the South; he had moved from Malta into Greece.

Between 1824 and 1827 Humphry revisited Wales, the Lake District, Ireland, and Scotland. In July 1825 Lockhart wrote of him in a postscript to his wife, 'This morning met Sir Humphry Davy in the highest glee, and a spick and span new *white* waistcoat, on his way to Connemara, etc. to fish. Broad white Beaver, lined with green as of yore.'

He was becoming apt to talk like a traveller. There is a slight sneer in Wordsworth's voice when he says that Sir Humphry was lavish in extolling the northern countries. Wordsworth had cooled towards him since the days when Davy was young and had twice paid visits to Grasmere. Then Wordsworth had thought him a most interesting man whose views were fixed on worthy objects. It was while both were staying at Lowther Castle as the guests of Lord Lonsdale that they were together for the last time. Of this meeting Wordsworth wrote from Ambleside to Sir John Stoddart in 1831:

I became acquainted with Sir Humphry Davy when he was a lecturer at the Royal Institution, and have since seen him frequently at his own house in London, occasionally at mine in the country and at Lord Lonsdale's, at Lowther, where I have been under the same roof with him for several days at a time. Of his scientific achievements I am altogether an incompetent judge; nor

[213]

did he talk upon those subjects except with those who had made them their study. His conversation was very entertaining, for he had seen much, and he was naturally a very eloquent person. The most interesting day I passed with him was in this country. We left Patterdale in the morning, he, Sir Walter Scott, and myself and ascended to the top of Helvellyn together. Here Sir H. left us, and we all dined together at my little cottage in Grasmere, which you must remember so well. When I last saw him which was for several days at Lowther (I forget the year), though he was apparently as lively as ever in conversation, his constitution was clearly giving way; he shrank from his ordinary exercises of fishing and on the moors. I was much concerned to notice this, and feared some unlucky result. There were points of sympathy between us, but fewer than you might perhaps expect. His scientific pursuits had hurried his mind into a course where I could not follow him; and had diverted it in proportion from objects with which I was best acquainted.

In 1840 Wordsworth, after reading the three-volume Life of the Quaker philanthropist William Allen, his own friend and Davy's, contrasted the two in a way unfavourable to Davy, saying that with all his intellectual power and extensive knowledge Davy was a sensualist, and a slave of rank and worldly station. 'I knew him well and it grieved me much to see that a man so endowed could not pass his life in a higher moral sphere.'

In 1826, on returning to London, Davy complained to Ayrton Paris of palpitation of the heart, and of an affection of the throat which made him fear the disease of which his father had died at the age of forty-eight. To Kitty he wrote of the difficulty of letter-writing because of rheumatism in his hand and arm; the pain, he thought, would not leave him till the weather changed. In a letter to Kitty dated July 2nd, 1826, he expressed his grief and alarm at the news of his mother's illness; but how little he feared that her sickness would be mortal is clear from the fact that he planned, on John's expected return to England, to go with him to Penzance for a family reunion.

The news of his mother's death – she had recovered from her first attack of illness, but she died suddenly in September after a second seizure – greatly distressed him. When he made Chemicus, in *Consolations in Travel*, say, 'That which I regarded most tenderly was in the grave', he was referring to his own loss. The virtue ascribed to Grace Davy by those who knew her well was the Christian virtue of meekness. One feels that the words written on the tablet in Ludgvan Church to the memory of this fine Cornishwoman imply, under the conventional phrasing, no mere lip-service.

The memorial was set up after Humphry's death, to his memory and to the memory of his parents, by John and the three sisters. Humphry was affectionate by nature, and his lack of domestic happiness in Park Street must have pointed a contrast with the home life of his boyhood. In the autumn of 1823 he wrote to John – after admitting that he was suffering from an irritable and diseased state of the liver – 'To add to my annoyances, I find my house, as usual after the arrangements made by the mistress of it, without female servants; but in this world we have to suffer and bear, and from Socrates down to humble mortals, domestic discomfort seems a sort of philosophical fate.' To John again, on hearing that his young brother was thinking of marriage, he wrote in January 1824, 'Upon points of affection it is only for the parties themselves to form just opinions of what is really necessary to ensure the felicity of the marriage state. Riches appear to me not at all necessary, but competence I think is; and after this, more depends upon the *temper* of the individual, than upon personal, or even intellectual circumstances. The finest spirits, the most exquisite wines, the nectars and ambrosias of modern tables, will all be spoilt by a few drops of bitter extract, and a bad temper has the same effect on life, which is made up not of great sacrifices or duties, but of little things, in which smiles and kindness, and small obligations given habitually, are what win and preserve the heart, and secure comfort.'

Davy's threatened breakdown in health became a reality in 1826, in the year of Scott's financial disaster. When, in 1825, Lockhart had met Sir Humphry setting off for Ireland to fish 'in the highest glee', Scott, too, to the casual observer, might seem to be in great felicity, courted and honoured by all, though for months, and even years, Ballantyne and Constable had been 'fluctuating between wild hope and savage despair'. By the beginning of 1826 Scott had faced the catastrophe which ruined him, and was displaying those qualities of indomitable courage and hope in desolation which make of his own life a tale as heroic as any he ever penned in fancy. His journal, his letters, and Lockhart's narrative move the heart to love and sympathy and admiration. His letter to Lady Davy, written in reply to a letter of concern from her, show both his own endurance and his hatred of all 'poor-manning'. While answering very civilly to Lady Davy who was, after all, related to him, and who had been very kind to his daughter Sophia, and to Lockhart his son-in-law when they settled in London, he yet confided to Lady Abercorn how much he hated that the London fashionable hostess should be taking thought for an old literary lion like himself. To his cousin he wrote on February 6th, 1826:

My dear Lady Davy,

A very few minutes since, I received your kind letter, and answer it in all frankness,—and, in Iago's words, 'I am *hurt*, ma'am, but not killed' — nor even kilt. I have made so much by literature, that even should this loss fall in its whole extent, and we now make preparations for the worst, it will not break, and has not broken my sleep. If I have good luck, I may be as rich again as ever; if not, I shall have still far more than many of the most deserving people in Britain — soldiers, sailors, statesmen, or men of literature.

I am much obliged to you for your kindness to Sophia, who has tact, and great truth of character, I believe. She will wish to take her company, as the scandal said ladies liked their wine, little and good; and I need not say I shall be greatly obliged by your continued notice of one you have known for a long time. I am, between ourselves, afraid for the little boy [John Hugh, the little son of Sophia and Lockhart]; he is terribly delicate in constitution and so twined about the parents' hearts, that — But it is needless croaking; what is written on our foreheads at our birth shall be accomplished. So far I am a good Moslem.

Lockhart is, I think, in his own line, and therefore I do not regret his absence; though, in our present arrangement, as my wife and Ann propose to remain all the year round at Abbotsford, I shall be solitary enough in my lodgings. But I always loved being a bear and sucking my paws in solitude, better than being a lion and ramping for the amusement of others; and as I propose to slam the door in the face of all and sundry for these three years to come, and neither eat nor give to eat, I shall come forth bearish enough, should I live to make another avatar.

Seriously, I intend to receive nobody, old and intimate friends excepted, at Abbotsford this season, for it cost me much more in time than otherwise.

The Beginning of Illness: Salmonia

I beg my kindest compliments to Sir Humphry, and tell him Ill
Luck, that direful chemist never put into his crucible a more indissoluble piece of stuff than your affectionate cousin and sincere
well-wisher,

WALTER SCOTT.

Before the year was out Sir Humphry, too, had been placed in the
crucible of Ill Luck. Like Sir Walter, he proved a pretty indissoluble
piece of stuff. The struggle he made against broken health gives a not
dissimilar impression of a valiant spirit. On the occasion of his Discourse delivered to the Royal Society at the anniversary meeting on
St Andrew's Day, 1826, he spoke with such effort that the drops of
sweat ran down his face, and those who were near him feared the kind
of apoplectic seizure which destroyed so many of his great contemporaries. For the seventh time he was elected President of the Society;
but he was so much exhausted that he was unable to attend the usual
dinner.

John Davy, who had returned to England, and who attended this
meeting of the Royal Society, went with his brother to Mr Watt
Russel's house in Northamptonshire – Mrs Watt Russel was the
daughter of Davy's old friend James Watt – and then the brothers
separated, John going into Monmouth and Humphry into Sussex
where, on a visit to Lord Gage at Firle, he was seized with the illness
which led to his first paralytic attack. He managed to reach his house
in Park Street to which John was summoned from Monmouth. He was
placed under the care of Dr Babington and Dr Holland, afterwards Sir
Henry Holland. The paralysis had affected his right side, but his
doctors hoped for a complete recovery. In due course they advised
a journey to Italy, partly because of the prospect of an earlier, warmer
spring, and partly in order that Davy might be free from the dis-
satisfactions and vexations attendant on his office as President of the
Royal Society. John Davy accompanied Humphry on this journey.

It was made through bitter weather, and began in a snow-storm.
In Italy, where they hoped they would have taken leave of winter,
they were disappointed. Snow through the whole of Lombardy lay
deeper than in the passes of the Alps. They stopped some days at
Bologna and when they reached Ravenna, in the first week of March,
the snow was melting on the roofs of the houses, and was to be seen in
the ditches some days later. Davy had chosen Ravenna for its promise

of solitude and repose; for its climate, its beauty, and its associations both historical and personal. Writing to Poole he said, 'Here Dante composed his divine works; here Byron wrote some of his best and most moral (if such a name can be applied) poems.'

John Davy had to return to his medical duties in the Ionian Islands, but Humphry was not without a friend. By the kindness of Monsignor Spada, Governor of the Province, he was lodged in the episcopal palace. As Spada was under the same roof, and much confined to the house because of his duties as governor, he and his guest saw enough of each other to relieve Davy of loneliness while not taxing too much his strength. Davy spoke always of Spada in terms of endearment and gratitude as an amiable and most enlightened companion whose conversation was like sunshine in the wintry state of his mind. He re-read at this time Byron's poems, of which he had procured at Calais a convenient travelling copy. He resumed, too, his acquaintance with the Countess Guiccioli who, when he first arrived, was living at Ravenna.

Davy never wearied of the *pineta*, the great stretches of pine groves, and he longed for Poole to see them. He said Poole must know the trees from Claude Lorrain's landscapes; he must imagine a circle of twenty miles of great fan-shaped pines, green sunny lawns, and little knolls of underwood, with large junipers of the Adriatic in front and the Apennines still covered with snow behind. In general, Davy writes uncommonly little about flowers; but in the Ravenna spring, the violets especially rejoiced him; there were so many they made the grass look purple. It is amusing to see how his descriptive prose, plain when writing to Poole, is more ornate when intended for Lady Davy.

He stayed in Ravenna until it was necessary to seek the cooler weather of the mountains. The journey northward was begun in the second week of April; he remained in various parts of Illyria and Austria until the beginning of October. From Salzburg, on July 1st, he wrote to Davies Gilbert resigning the Presidency of the Royal Society. The letter reveals his condition, his constant interest in the Royal Society and the hope that Robert Peel, who had been his supporter in many an important development in relation to the Society, might succeed him as President.

To Lady Davy he wrote of his enforced resignation and of his feelings towards the Society. Had he perfectly recovered, his decision, under the auspices of a new and more enlightened government, might, he thought, have been different; but his state of health rendered the relinquishing of the chair absolutely necessary. He said there was no period of his life at which he looked back with more real satisfaction than the six years of labour for the interests of the Society. He never was and never could be, he thought, unpopular with the leading and active members as six unanimous elections proved. 'But', he continued, 'because I did not choose the Society to be a tool of Mr —'s journal jobs, and resisted the admission of improper members, I had some enemies who were listened to and encouraged from Lady —'s chair. I shall not name them, but, as Lord Byron said, "my curse shall be forgiveness".' When he had decided to return to England he wrote to his wife, 'I think you will find me altered in many things – with a heart still alive to value and reply to kindness, and a disposition to recur to the brighter moments of my existence of fifteen years ago, and with a feeling that though a burnt-out flame can never be rekindled a smothered one may be.'

Lady Davy herself had not been well. In his letters Davy teased her about her 'Pope in medicine', Mr Clarke, later Sir Charles Clarke. In 1843 Sir Charles was to cure her cough by prescribing 'a morning draft of a *small* quantity of *old* rum in new milk'; later he prescribed curaçao in hock instead of medicine, so she had cause to think him infallible. Although Lord Dudley had found her, in his weakness and melancholy, a delightful, cheerful, and stirring travelling companion on a little tour to Fonthill – no scandal in the world, he had assured the Bishop of Llandaff – she was really a woman the least fitted to take steady care of an invalid. Yet to the house in Park Street Davy returned in the second week in October. In *Consolations in Travel*, under a thin disguise, he contrasts his youthful joy on entering London with his later disenchantment:

> In my youth, and through the prime of manhood, I never entered London without feelings of pleasure and hope. It was to me as the grand theatre of intellectual activity, the field of every species of enterprise and exertion, the metropolis of the world of business, thought and action. There I was sure to find the friends and companions of my youth, to hear the voice of encouragement and praise. There society of the most refined kind offered daily its banquets to the mind, with such variety, that satiety had no place in them, and

new objects of interest and ambition were constantly exciting attention either in politics, literature, or science.

I now entered this great city in a very different tone of mind, one of settled melancholy, not merely produced by the mournful event which recalled me to my country, but owing likewise to an entire change in the condition of my physical, moral and intellectual being. My health was gone, my ambition was satisfied, I was no longer excited by the desire of distinction; what I regarded most tenderly was in the grave, and to take a metaphor, derived from the change produced by time on the juice of the grape, my cup of life was no longer sparkling, sweet and effervescent; – it had lost its sweetness without losing its power.

As soon as it became clear that Davy could not resume his former life in London the plan of buying a house or small estate in the West, proposed as early as 1824, was again pursued. He wrote to Poole that he had thought of Minehead, Ilfracombe, Lynmouth, and Penzance; the objection to Penzance being the fear of too much society. He wanted only a very little of anything, he said – a little riding, a very little shooting, a little quiet enjoyment with friends, some facilities for procuring books, and a temperate situation.

Poole received him as a guest in November and December of 1827 and introduced him to young William Baker of Taunton.[1] William Baker was one of the many who owed encouragement to Poole. Mrs Sandford in her biography of Poole tells how, twenty years previously, Mr Baker senior, a currier by trade, saw Poole's book-room and said, 'My son Bill, sir, would love to see these books.' Poole said he would be glad to see Bill at any time, and the next morning, a Sunday morning, Bill arrived, walking from Taunton before breakfast. That was in the summer of 1807 when Coleridge was staying at Stowey. During the ensuing twenty years William Baker had been Poole's frequent companion; he had read much and had, above all, observed. Like Davy, he was in love with the exquisite make of the creatures of the world, and curious to investigate their wonders. His drawings were shown to Davy by Poole, Davy asked to meet him, encouraged him in his investigation of the generation of eels, and in observing the differences and similarities between eels and congers. Congers and the problems related to them had been in Davy's mind since the days when

---

[1] *A Brief Memoir . . . of William Baker*, by John Bowen, Taunton, 1854.

as a boy he had seen the Cornish congers brought into Penzance harbour and had talked with the fishermen about them.

He could follow his old favourite subject of natural history without the strain likely to arise from too severe application. But Poole shows in the letter he wrote Ayrton Paris, giving particulars of this visit, how Davy's inherent love of the laboratory persisted. At Fyne Court, Broomfield, in the Quantocks, lived Mr Andrew Crosse,[1] whose experiments in electro-crystallization had gained him the reputation in the neighbourhood of being a wizard-squire. The delicacy and precision of his arrangements made Kenyon say of him that he manipulated like a cat. Like Davy, he could retire into the recesses of his own mind and plan scientific arrangements and feel a poet's elation at their success. 'Often have I,' he wrote, 'when in perfect solitude, sprung up in a burst of schoolboy delight at the instant of a successful termination of a tremblingly anticipated result. Not all the applause of the world could repay a real lover of science for the loss of such a moment as this.' He had written verse in his younger days; W. S. Landor had lamented his defection from the Muse. Landor wrote:

> Although with earth and heaven you deal
> As equal, and without appeal,
> And bring beneath your ancient roof
> Records of all they do, and proof,
> No right have you, sequestered Crosse,
> To make the Muses weep your loss.
> A poet were you long before
> Gems from the struggling air you tore,
> And bade the far off flashes play
> Along your woods, to light your way . . .
> Southey, the pure of soul, is mute!
> Hoarse whistles Wordsworth's watery flute,
> Which mourn'd with loud indignant strains
> The famisht Black in Corsic chains . . .

It was Crosse who, at this time, because of his admiration for Davy, offered him his house during his illness. If this could have been arranged Lady Davy might have joined him there. Davy went with Poole to Broomfield to visit Mr Crosse, to see Fyne Court, which

---

[1] *Memoirs of Andrew Crosse the Electrician,* by Camelia A. H. Crosse, 1857.

contained a laboratory well equipped for the study of electricity and chemistry. Andrew Crosse said, 'Never shall I forget seeing Davy's fine melancholy eyes brighten up; as he looked at the furnaces for a few moments he seemed himself again, the languor of disease had fled, and his old activity was expressed in every look and action. But he was passing away, it was the beginning of the end.' Poole, too, was much impressed by this visit to the laboratory. He wrote of Davy, 'He threw his eyes round the room, which brightened in the action – a glow came over his countenance, and he appeared himself twenty years ago. He was surprised and delighted, and seemed to say, "This is the beloved theatre of my glory!" I said, "You are pleased?" He shook his head, and smiled.' He did not take Fyne Court.

During November and December, while he was seeking to restore his strength in the quiet ways of Nether Stowey, he began a book which he completed in 1828 and entitled, *Salmonia, or Days of Fly-Fishing*. He enlarged the title by describing the book as a series of conversations, with some account of the habits of fishes belonging to the genus *Salmo*. The speakers are Halieus, who has a good deal in common with Dr Babington – the William Babington to whom the book is dedicated, 'In remembrance of some delightful days passed in his society, and in gratitude for an uninterrupted friendship of a quarter of a century'; Physicus, who is nearest to Davy himself; Poietes, who moralizes, and Ornither who is full of information. Halieus is a contrast to Physicus. The tones of his mind are quiet. It has been so little scorched by sunshine, and so little shaken by the wind, that its climate is that fabled of the Hesperides, 'Where the same trees produced at once buds, leaves, blossoms and fruits.' Physicus desired for himself his friend's quietude; but on the whole the personages of the conversations have little individuality. They are vehicles to convey the discursive amplitude of Davy's curious knowledge, his love of rivers and meadows, and his pleasure in combining sport with natural history. Davy loved Izaak Walton, but he could not emulate him. Walton's 'study to be quiet' is not mere advice; the quietude is in his prose; whereas Davy's *Salmonia* is nearer in some ways to the nineteenth-century educational 'dialogues' than to *The Compleat Angler*. One almost catches the tone of Mrs Markham:

> 'Pray, Mamma, what was a Druid?'
> 'A Druid, my dear, was a kind of clergyman.'

The Beginning of Illness: Salmonia

But although the tone is didactic, and little art is shown in the general management of *Salmonia*, yet its very clumsiness endears the book as one gets to know it better. As the four interlocutors open their minds on different subjects, and pass from fact to fact or from idea to idea, one acknowledges oneself in good company. The very idea of people talking together for long on set themes is undramatic, though a master like Dryden could make the minds of his speakers take active fire from one another. Davy studied Dryden's prose, and even before leaving Penzance he had read Robert Boyle's *The Sceptical Chymist*, which is in dialogue form. Davy occasionally manages lively communication, as when Ornither, after showing why the auguries of the ancients were a good deal founded upon the observation of the instincts of birds, passes on to the subject of popular superstitions. In the course of the conversation Cornish readers will be interested to find Davy's explanation of the Bucca. He says that in the West of England, half a century before, a particular hollow noise on the sea coast was referred to a spirit or goblin, called Bucca, and was supposed to foretell a shipwreck. 'The philosopher knows that sound travels much faster than currents of air; and the sound always foretold the approach of a very heavy storm, which seldom takes place on that wild and rocky coast, surrounded as it is by the Atlantic, without a shipwreck on some part of its extensive shores.' So to hear Bucca was unlucky.

The anglers discourse for a time on popular superstitions and then pass to a consideration of the profound relations of events. Halieus remarks, 'I have in life met with a few things which I found it impossible to explain, either by chance, coincidence, or by natural connections; and I have known minds of a very superior class affected by them, – persons in the habit of reasoning deeply and profoundly.' Physicus replies, 'In my opinion profound minds are the most likely to think lightly of the resources of human reason: it is the pert, superficial thinker, who is generally strongest in every kind of unbelief. The deep philosopher sees chains of causes and effects so wonderfully and strangely linked together, that he is usually the last person to decide upon the impossibility of any two series of events being independent of each other; and in science, so many natural miracles, as it were, have been brought to light, – such as the fall of stones from meteors in the atmosphere, the disarming a thunder-cloud by a metallic point, the production of fire from ice, by a metal white as silver, and referring certain laws of motion of the sea to the moon, – that the

[223]

physical inquirer is seldom disposed to assert, confidently, on any abstruse subjects belonging to the natural order of things, and still less so on those relating to the more mysterious relations of moral events, and intellectual natures.'

This declaration of one article of Davy's faith is made at the close of the Sixth Day's fishing. *Salmonia* is divided into nine Days, very long Days, often with part of the Night thrown in. The scenes are widely different and very far apart. In *Salmonia* we do not actually travel; it is taken for granted that the four friends meet by the various rivers and lakes – the Colne; Loch Maree, in Ross-shire; Leintwardine, near Ludlow; or the Fall of the Traun in Upper Austria.

On the First Day the scene is in London, and the friends – Halieus, Physicus and Poietes – are discovered not catching but eating trout taken in the Wandle. They discuss fishing in general, remembering those who despise fishing and those who love to fish. They recall Dr Johnson's dictum that angling is an amusement with a stick and a string, a worm at one end and a fool at the other; and Byron's description of Izaak Walton as 'a quaint old cruel coxcomb'. Physicus is made to condemn angling – he is later to be an initiate – Halieus to defend it. In the course of his defence of Walton against the charge of cruelty, Halieus quotes a poem addressed to the memory of the old angler written by a lady 'long distinguished at court for pre-eminent beauty and grace'. The poem is signed C. C. 1812. It has never, as far as I know, found a place in anthologies, and yet it has charm of movement, and a certain fragrance in the period diction, as of petals long laid up. C. C. is said to have written the verses into the copy of *The Compleat Angler* which Halieus carried about with him. It will be remembered that Davy had wooed Mrs Apreece with a copy of old Izaak; but the verses are too delicate to have been made by her, and the initials are not hers.

The lady addresses Izaak Walton:

> Albeit, gentle Angler, I
>   Delight not in thy trade,
> Yet in thy pages there doth lie
> So much of quaint simplicity,
>   So much of mind,
>   Of such good kind,
>   That none need be afraid,

Caught by thy cunning bait, this book,
To be ensnared on thy hook.

Gladly from thee, I'm lured to bear
    With things that seemed most vile before,
For thou didst on poor subjects rear
Matter the wisest sage might hear.
      And with a grace,
        That doth efface
More laboured works, thy simple lore
Can teach us that thy skilful *lines*
More than the scaly brood *confines*.

Our hearts and senses, too, we see,
    Rise quickly at thy master hand,
And, ready to be caught by thee,
Are lured to virtue willingly.
      Content and peace,
        With health and ease
Walk by thy side. At thy command
We bid adieu to worldly care,
And joy in gifts that all may share.

Gladly with thee I pace along,
    And of sweet fancies dream;
Waiting till some inspired song,
Within my memory cherished long,
      Comes fairer forth,
        With more of worth;
Because that time upon his stream
Feathers and chaff will bear away,
But give to gems a brighter ray.

Much space on the Sixth and Seventh Days is devoted to the instincts of fishes, to the habits of the grayling, and to the generation and migration of eels. The Eighth Day is concerned especially with the natural history of insects, and the Ninth is more discursive than all the others, the conversation passing rapidly from the hucho to the sea-serpent and from mermaids to education. The tale of the mermaid of

Caithness illustrates the antiquity of mythical creatures in sea and loch. Halieus explodes the legend as current in his day by adducing a prosaic explanation:

> The mermaid of Caithness, was certainly a *gentleman*, who happened to be travelling on that wild shore, and who was seen bathing by some young ladies at so great distance, that not only *genus*, but gender, was mistaken. I am acquainted with him and have had the story from his own mouth. He was a young man, fond of geological pursuits; and one day in the middle of August, having fatigued and heated himself by climbing a rock to examine a particular appearance of granite, he gave his clothes to his Highland guide, who was taking care of his pony, and descended to the sea. The sun was just setting, and he amused himself for some time by swimming from rock to rock, and having unclipped hair and no cap, he sometimes threw aside his locks, and wrung the water from them on the rocks. He happened the year after to be at Harrogate, and was sitting at table with two young ladies from Caithness, who were relating to a wondering audience the story of the mermaid they had seen, which had already been published in the newspapers: they had described her, as she usually is described by poets, as a beautiful animal with fair skin, and long green hair. The young gentleman took the liberty, as most of the rest of the company did, to put a few questions to the elder of the two ladies – such as, on what day, and precisely where, this singular phenomenon had appeared. She had noted down, not merely the day, but the hour and minute, and produced a map of the place. Our bather referred to his journal, and showed that a human animal was swimming in the very spot at that very time, who had some of the characters ascribed to the mermaid, but who had no claim to others, particularly the green hair and fish's tail; but being rather sallow of face, was glad to have such testimony to the colour of his body beneath his garments.

Hands and a finny tail were, said Halieus, entirely contrary to the analogy of Nature, and so he disbelieved in the mermaid on philosophical principles. He says that the dugong and the manatee are the only animals combining the functions of mammalia with some of the characters of fishes that can be imagined even as a link. Earlier, Davy had seemed distinctly to anticipate in a crude form Charles Darwin's theory. The passage occurs in a speech made by him in 1822 when he presented the Copley medal to Doctor Buckland. Sound geology had dispersed, he said, 'the eternal order of one simple system'. Geology had discerned a progressive development in the appearance of living beings upon the earth. It discerned 'organised germs passing, in the immensity of time, through the different stages of improvement, rising from fishes through mermaids, quadrupeds, and apes: and at last perfect in man'.

In *Salmonia* there is an interesting reference to Erasmus Darwin, author of *Loves of the Plants* which provoked Canning's *Loves of the Triangles*. Physicus teases Halieus by comparing his doctrine of the hereditary transmission of acquired characteristics to Dr Darwin's. Physicus says, 'I am somewhat amused at your idea of the change produced in the species of trout, by the formation of particular characters, by particular accidents and their hereditary transmission. It reminds me of the ingenious, but somewhat unsound views of Darwin [i.e. Erasmus Darwin] on the same subject.' Halieus replies, 'I will not allow you to assimilate my views to those of an author, who, however ingenious, is far too speculative; whose poetry has always appeared to me weak philosophy, and his philosophy indifferent poetry: and to whom I have been accustomed to apply Blumenbach's saying, "that there were many things new, and many things true, in his doctrines; but that which was new was not true, and what was true, was not new".'

When the time for the fishermen finally to take leave of one another approaches, Halieus suggests that they may meet again in England in the winter, and hopes they will confess that the time bestowed on angling has not been thrown away. He thinks perhaps the most important principle in life is to have a pursuit – a useful one if possible, and at all events an innocent one. The peculiar effect of angling on himself, he says, is to bring him back to early times and feelings, and to create again the hopes and happiness of youthful days.

This remark calls out the most poignant expression we have of what Davy felt to be the contrast between his own early and later days. In the person of Physicus he tells Halieus that he too has felt something like what has been described; but he knows that the freshness and glory of his youth can never return. He says:

I felt something like what you described, and were I convinced that, in the cultivation of the amusement, these feelings would increase,

I would devote myself to it with passion; but, I fear, in my case that is impossible. Ah! could I recover anything like that freshness of mind, which I possessed at twenty-five, and which, like the dew of the dawning morning, covered all objects and nourished all things that grew, and in which they were more beautiful even than in mid-day sunshine, – what would I not give! – All that I have gained in an active and not unprofitable life. How well I remember that delightful season, when, full of power, I sought for power in others; and power was sympathy, and sympathy power; when the dead and the unknown, the great of other ages and distant places, were made, by the force of the imagination, my companions and friends; when every voice seemed one of praise and love; when every flower had the bloom and odour of the rose; and every spray or plant seemed either the poet's laurel, or the civic oak – which appeared to offer themselves as wreaths to adorn my throbbing brow. But, alas! this cannot be; and even you cannot have *two springs* in life – though I have no doubt you have fishing days, in which the feelings of youth return, and that your autumn has a more *vernal* character than mine.

18 · 'Consolations in Travel'

Salmonia was published by John Murray. Letters from Lady Davy to the publisher, letters written in her Italianate hand, which have been preserved at 50 Albemarle Street, show her solicitous to prevent her husband's health and spirits from being affected by delays or by vexatious notices. She requested John Murray to send him the *Quarterly Review* but not to allude to *Blackwood's Magazine* – 'It is needless, and though I hope Sir Humphry would not mind what I suppose it to be, malignant and unfeeling criticism, I do not wish his mind attracted even by mere curiosity to any disagreeable reflections. I think he will not *seek* Blackwood's, and I am for the credit of taste and good feeling inclined to hope it will not readily *find* him.' Her letters to Murray reveal something of the kind and genuine Scot in her nature – and her preoccupation with her health. In one letter she wrote, 'I am very unwell. I hope Mrs Murray's appetite is better than my own. I send her a pheasant which is a wholesome dinner . . . I add 5 Portugal onions which your London vitiated taste may prefer to Scotch leeks, which Mrs Murray with myself will not however have despised I daresay.'

Salmonia had a welcome from readers, though Christopher North ridiculed it, and the remarks of Poietes on the dangers of education forced on people produced a virulent attack on Davy from the President of the Mechanics' Institute, who accused him of conceit, pride, and arrogance. Davy, who always emphasized the necessity for a liberal education, was the natural critic of Brougham's 'steam intellect societies'.

Mrs Somerville praised the book, and Scott's charming notice of it in the *Quarterly Review* pleased Davy so much that he wrote to Lady Davy that it had almost rekindled his love of praise, and given him again 'heroic thoughts'. His mind was turning strongly towards the ideas of his youth, towards poetry and religion. He hoped that the final book he wrote, *Consolations in Travel* or *The Last Days of a Philosopher*

[229]

(published posthumously in January 1830), might be considered as a kind of poetry, though in it he attempted no verse. Like *Salmonia*, *Consolations in Travel* is the work of a man of insight whose powers of co-ordination were failing; yet this book affects the mind more strongly than many a better composed piece. Lady Davy told John Murray that it was to her a 'painfully interesting volume'.

Almost certainly Davy knew the two volumes of Landor's *Imaginary Conversations* which had appeared in 1824. Landor was Southey's friend. Davy's prose has not the music of Landor's; but when the matter prompts he reaches at moments a grave beauty. There are utterances in *The Proteus, or Immortality* and *Pola, or Time* not unworthy to stand by the prose of *Imaginary Conversations*. Davy comes to no conclusion: the form of argument enables him always to put the case for the opposition as he proceeds.

Consolations in Travel consists of six Dialogues with a fragment of a seventh. The speakers – Ambrosio, Onuphrio, Eubathes, the Unknown, and Philalethes – have interesting prototypes in the actual life of their time. If we think of them as Spada, William Lamb, Wollaston, and Davy himself under a couple of disguises, the whole becomes more alive – but then, Davy would not have been so free to use contrarieties if he had used real names, and in these contrarieties the inner pulse of the work beats.

Ambrosio has a good deal in common with Monsignor Spada, Vice-Legate of Ravenna, and governor of the province, by whose kindness Davy was lodged in the apostolical palace during his stay at Ravenna, and whose conversation and skill at écarté relieved the tedium of his last days in Rome. In the Dialogues he is described as a Roman Catholic of the most liberal school, a man of highly cultivated taste, great classical erudition, and minute historical knowledge. It is significant that in the year 1828–9 Davy should have chosen a Roman Catholic as his representative upholder of the Christian faith. Ambrosio speaks no theology; he is made to argue for man as an accountable and immortal being – 'know thyself first immortal', as Chaucer phrased it.

In contrast with Ambrosio, Onuphrio expresses lightly his teasing, indolent, unzealous view. There is nothing to prove that Davy had William Lamb in mind and yet, as one reads Lord David Cecil's biography of the young Melbourne, one is irresistibly reminded of Davy's imagined Onuphrio.

'Consolations in Travel'

Philalethes comes nearest to Davy himself. He is conscious of his own ignorance, but his delight is in inquiry; to him contemplation constitutes a pure and abstract pleasure. Far from making his mind hostile to religion, science and his experience of life have made him yearn for it.

The mind sufficient to itself in the exercise of its reasonable functions, the mind insensible to poetry and indifferent if not actually hostile to religion is represented by Eubathes, who has some of the characteristics of Dr Wollaston: 'He [Eubathes] was a man of a very powerful and acute understanding but had less of the poetical temperament than any person whom I had ever known with similar vivacity of mind. He was a severe thinker, with great variety of information, an excellent physiologist, and an accomplished naturalist. In his reasoning, he adopted the precision of a geometer, and was always upon his guard against the influence of imagination.'

Through these characters, in *Consolations in Travel*, Davy has bequeathed us a book in which there is much dead matter, which is without a firm centre, and which nevertheless has an intrinsic interest. It shows us a mind struggling to see beyond the prejudices and ignorances of its mortal environment. In an age turning away from romantic dreams of liberty towards utilitarianism, and towards a form of Christianity which seemed itself the instrument of materialistic comfort, Davy saw man as a spirit. He grasped firmly the notion that 'surely there is a piece of divinity in us, which was before the elements and owes no homage to the sun'. He wrote a mixture of description, argument, speculative science, autobiography, history, and weak fiction. But he had a vision, and an assurance, as well as adumbrations of future conquests of knowledge.

The scene of the vision is the Colosseum on a fine October afternoon. After some talk between the friends Davy desires to be left alone in the solitude of the ruins. As night approaches he expresses the romantic melancholy which the sensibility of the time demanded, but which is also personal to him. His mood of sadness deepens, but only to give place to a vision, that vision of progress which was to lighten the liberal imagination of his successors, and which, indeed, was to become so bright as to blind them to much that Davy himself was never blind to. Absorbed into a flood of splendour, Davy hears a melodious voice which tells him that he, like his brethren, is entirely ignorant of everything belonging to man, the world he inhabits, his future destiny, and

[231]

the scheme of the universe; and yet he has the folly to believe he is acquainted with the past, the present, and the future. There are, he is told, intelligences as superior in power and intellect to man as man is above the meanest and feeblest reptile that crawls beneath his feet. The butterfly does not transport with it into the air the organs and the appetites of the crawling worm from which it sprang. So the monad or spirit leaves behind it our earthly system of organized dust. It drinks intellectual light from a purer source approaching nearer to the infinite divine mind. To one whom he names the Genius, Davy is bidden to yield his mind wholly. Then he is borne through space with unsealed eyes. The Genius lectures. The result is rather like the result when Yeats takes down his wife's automatic writing – a medley of remembered philosophy, insight, and imagination. Davy's vision of society is of an unsteady, checked, terrible, broken but genuine progress. The glory and greatness of the military genius of a nation passes away; its arts and institutions arise in their vigour in another state of society. He has no sympathy with the idea of the 'noble savage'. He sees that in histories of the world all the great changes of nations are confounded with changes in their dynasties, and events are usually referred either to sovereigns, chiefs, heroes, or their armies. But governments depend far more than is generally supposed upon the opinion of the people and the spirit of the age and nation. It sometimes happens that a glorious mind possesses supreme power and rises superior to the age in which it is born. Among others, Davy cites King Alfred, of whom he was a great admirer. But he shows that more often benefits have come from the despised and neglected – men who have sacrificed all their common enjoyments and all their privileges as citizens to the pure search for truth, or to the practical benefiting of their fellows. He cites Anaxagoras, Archimedes, Roger Bacon, and Galileo as among those who in their deaths and their imprisonments offer instances of the kind. He finds nothing more striking in history than what appears to be the ingratitude of men towards their greatest benefactors. His confidence in the future of society is based on the evidence that great and real improvements are perpetuated. 'A new period of society may send armies from the shores of the Baltic to those of the Euxine and the empire of the followers of Mahomet may be broken in pieces by a Northern people, and the dominion of the Britons in Asia may share the fate of that of Tamerlane or Genghis Khan; but the steamboat which ascends the Delaware or the St Lawrence will continue to be used, and will carry the civilization of an improved people into the wilds of Canada.' He sees the progress of society as bound up with the progress of physics, chemistry, and mechanics which had already produced results of a marvellous kind.

The vision which the Genius unfolds for the individual spirit is also one of progression. It is a philosophy which many poets have made familiar. Davy imagines spiritual natures as eternal. His vision of the intellectual love of God, in which wisdom consists, is Spinoza's.

In the Second Dialogue between Ambrosio, Onuphrio, and Philalethes, which is imagined to occur one evening on the summit of Vesuvius, Ambrosio teases Philalethes about his Vision, which he considers a sort of political epitome of his philosophical opinions – the dream being the mere web of imagination. Philalethes admits an admixture of fiction, but contends that parts of the vision came to him in actual dream. He says no absolutely new ideas are produced in sleep, but that most extraordinary combinations occur. Ambrosio reproaches him for ignoring revealed religion and himself confesses that he founds his Christianity rather upon the fitness of its doctrines, the manner of its spreading, the exalted nature of those men by whom it has been professed and examined, than upon historical evidences. The infinite can never be understood by the finite. Ambrosio thinks religion has nothing to do with the common order of events; it is a pure and divine instinct intended to give results to man which he cannot obtain by the common use of his reason; and which at first view often appear contradictory to it. Considered in the most extensive and profound relations, they are in fact in conformity with the most exalted intellectual knowledge, so that indeed the results of pure reason ultimately become the same with those of faith, – 'the tree of knowledge is grafted upon the tree of life, and that fruit which brought the fear of death into the world, budding on an immortal stock, becomes the fruit of the promise of immortality'. Ambrosio is pragmatical, and Onuphrio entices him into difficult admissions. The whole discussion, Onuphrio says, makes him philo-Christian, but he can neither understand nor embrace the views developed. Ambrosio abjures him to fix his powerful mind upon the harmony of the moral world as he has been long accustomed to do upon the order of the physical universe; then he would see the scheme of the eternal Intelligence developing itself in both equally. He must aid his contemplation with devotional feelings and mental prayer, and wait with humility for the light.

The Third Dialogue, which introduces the Unknown, shows the friends leaving Naples – as Davy remembered leaving it – at three in the morning to visit the remains of the temples at Paestum. Relays of horses have been provided and at half-past one o'clock they find themselves descending the hill of Eboli towards the plain. Davy says that were his existence to be prolonged through ten centuries he could never forget the pleasure he received when they alighted to eat under a pine tree and looked about them. It is the best piece of natural description in the book. The lines are clear, and the colours, and the sound of the breeze in the pines overhead, and the rattling of the cones.

It is when the travellers descend the hill and are examining the ruins that they meet the Unknown. The Unknown's person and adventures are the fruit of Davy's fancy, but his voice is the author's. He has Davy's eager curiosity, ranging through particular knowledge of travertine to general questions of geology and the origin of species. Onuphrio shows himself as recalcitrant to faith in new-fangled ideas as to those based on revelation. He tells the story of a deception practised by a geologist on some Italian peasants. The geologist was on Etna, and busily employed in making a collection of the lavas formed from the igneous currents of that mountain; the peasants were often troublesome to him, suspecting that he was searching for treasure. It occurred to him to make the following speech to them: 'I have been a great sinner in my youth, and, as a penance, I have made a vow to carry away with me pieces of every kind of stone found upon the mountain; permit me quietly to perform my pious duty that I may receive absolution for my sins.' The speech produced the desired effect. The peasants shouted, 'The holy man, the saint!' and gave him every assistance in their power to enable him to carry off his burthen, and he made his ample collections with the utmost security and in the most agreeable manner.

The Stranger says he does not approve of pious frauds even for philosophical purposes. But both he and Ambrosio seem to base their Christianity on feeling, and on the desirability of inculcating good feeling, rather than on such intellectual reasoning as would have satisfied St Thomas Aquinas.

The Fourth, Fifth, and Sixth Dialogues are the most interesting, although one misses the presence of Onuphrio. For him Eubathes (Dr Wollaston) is substituted. The scene is the Alpine country of Austria; the period later than for the first three Dialogues. The references to

Davy's own illness and the loss of his mother are very thinly veiled; the settled melancholy which has seized on him is mournfully expressed. He is obsessed with the brevity of the lives of ardent statesmen, poets, and philosophers. The English he sees as a nation pre-eminently active, following their objects with force, fire, and constancy. Their climate necessitates activity; for repose he seeks the Alpine country of Austria and it is in this country, loved for its sparkle, change, grandeur, and good fishing, that he sets the Fourth Dialogue.

Davy, like Dr Johnson in *Rasselas*, is weakest when inventing incident. The precipitation of Philalethes in his boat over a waterfall, and the landing of his body, like a great fish, by the gaff of the Unknown must have made Scott smile. But when Philalethes, Eubathes, and the Unknown begin to talk, the book, like the sometime dead-looking philosopher, picks up. They talk of the *proteus anguinus* (which the *Oxford English Dictionary* defines as 'a genus of tailed amphibians with persistent gills, having four short slender legs and a long eel-like body'). The examination of these amphibious creatures leads to a long discourse on respiration in which Davy was interested from his Bristol days, and passes to a consideration of the relation between organized matter and 'life' or 'vitality'. The discussion is too elaborate and closely knit to summarize, but some examples of its eloquence may be quoted. The Unknown is replying to Eubathes, who feels that the many new and extraordinary views in electricity may be applied to solve some of the mysterious and recondite phenomena of organized beings. He says, 'But the analogy is too remote and incorrect; the sources of life cannot be grasped by such machinery; to look for them in the power of electro-chemistry is seeking the living among the dead; – that which touches will not be felt, that which sees will not be visible, that which commands sensations will not be their subject.'

In this Dialogue of *The Proteus*, or *Immortality*, as elsewhere both in *Salmonia* and in *Consolations in Travel*, the great value of the form adopted is that it permits a man's opponent his say; it allows for the divided mind and the partial allegiance; it never tries to prevent the human mind from piercing a little further. The Unknown is driven to confess his entire ignorance of the causes of vitality. He says:

> I confess with the utmost humility my ignorance. I know there have been distinguished physiologists who have imagined that by

organization, powers not naturally possessed by matter were developed, and that sensibility was a property belonging to some unknown combination of unknown etherial elements. But such notions appear to me unphilosophical and the mere substitution of unknown words for unknown things. I can never believe that any division, or refinement, or subtilization, or juxtaposition, or arrangement of the particles of matter, can give to them sensibility; or that *intelligence* can result from combinations of insensate and brute atoms. I can as easily imagine that planets are moving by their will or design round the sun, or that a cannon-ball is reasoning in making its parabolic curve. The materialists have quoted a passage of Locke in favour of their doctrine, who seemed to doubt 'whether it might not have pleased God to bestow a power of thinking on matter'. But with the highest veneration for this great reasoner, the founder of modern philosophical logic, I think there is little of his usual strength of mind in this doubt. It appears to me that he might as well have asked, whether it might not have pleased God to make a house its own tenant . . .

The argument sways to and fro, but always it comes back to the same thing – that although it can be proved that a certain perfection of the machinery of the body is essential to the exercise of the powers of the mind, such reasoning does not prove that the machine is the mind. Without the eye there can be no sensations of vision, and without the brain there can be no recollected visible ideas; but neither the eye nor the brain can be considered as the percipient principle – each is but the instrument of a power that has nothing in common with them. We are as ignorant of the divine fire as a savage of the fire which moves a complicated piece of machinery. Natural and revealed religions are to man what instincts are to animals, prompting him to good, and awakening in him feelings of gratitude and divine love. But the wish that man might be guided in life and comforted at the hour of death by religious precepts does not make Davy shrink from urging man to push his understanding of the universe to the utmost limit. The device of dialogue enables him never to be treacherous to his master principle – that there is no impiety in seeking to understand the nature of the world though there may be impiety in the use made of the knowledge.

Understanding can, he thinks, be advanced by the study of chemistry, and the Fifth Dialogue is called *The Chemical Philosopher*. In this Dialogue the Unknown is the principal speaker and, after giving a brief outline of his life, he defends his favourite study against the attacks of Philalethes and Eubathes. Philalethes suggests that higher mathematics and pure physics would have offered more noble objects of contemplation and fields of discovery. They suggest that the results of the chemist are more humble, belonging principally to the apothecary's shop and the kitchen.

The Unknown meets them even on this teasing ground, and says that the beginning of the progress of Civilization is in the discovery of some of the useful arts through chemical and mechanical inventions; he says that science is nothing more than the refinement of common sense making use of facts already known to acquire new facts. Pressed for instances, and confronted with the suggestion that accident has had as much to do with discovery as genius, he warms to his subject, pouring out instance after instance of chemical research advancing processes useful to common life.

Eubathes then changes his ground. He says that it is the chemical manufacturers pursuing practice for gain who have done more for society than the chemical philosopher. He sneers at mediocre persons claiming the title of 'philosopher' (Coleridge says something very similar in one of his letters) because they have dissolved a few grains of chalk in an acid, or shown a very useless stone to contain certain known ingredients. The Unknown ignores the sneer. He contends that many processes could only have been arrived at after multiplied experiments based on most ingenious and profound views. Then, waiving all notions of utility and vulgar applications, he defends the study of chemistry as a pure delight in knowing and understanding the processes of nature, and in contemplating the order and harmony of the arrangements belonging to the terrestrial system of things: 'to produce, as it were, a microcosm in the laboratory of art, to measure and weigh the invisible atoms, which, by their motions and changes, according to laws impressed on them by the Divine Intelligence, constitute the universe of things. The true chemical philospher sees good in all the diversified forms of the external world. Whilst he investigates the operations of infinite power guided by infinite wisdom, all low prejudices, all mean superstitions disappear from his mind. He sees man an atom amidst atoms fixed upon a point in space; and yet modifying the laws that are around him by understanding them; and gaining, as it were, a kind of dominion over time and an empire in material space and exerting on a scale infinitely small, a power seeming a sort of shadow or reflection of a creative energy, and which entitles him to the distinction of being made in the image of God, and animated by a spark of the Divine mind.' Whilst chemical pursuits exalt the understanding, they do not depress the imagination or weaken genuine feeling; '. . . they keep alive that inextinguishable thirst after knowledge, which is one of the greatest characteristics of our nature; for every discovery opens a new field for investigation of facts – shows us the imperfection of our theories. It has justly been said that, the greater the circle of light, the greater the boundary of darkness by which it is surrounded. This strictly applies to chemical inquiries; and hence they are wonderfully suited to the progressive nature of the human intellect, which by its increasing efforts to acquire a higher kind of wisdom, and a state in which truth is fully and brightly revealed, seems as it were to demonstrate its birthright to immortality.'

He defines chemistry: 'Chemistry relates to those operations by which the intimate nature of bodies is changed, or by which they acquire new properties.' He also gives his idea of the kind of education a chemist needs, and the qualities necessary for discovery, or the advancement of science.

Davy's optimism depends on his faith in science and progress; his sadness from his perception of the decay of all things. His own body was in such a process of decay.

The tone of the Sixth Dialogue, named *Pola*, or *Time*, is the tone of the author of Ecclesiastes. Appropriately the dialogue is staged at Pola, whose harbour the friends enter in a felucca at sunset. They examine the monuments of imperial grandeur, and reflect on the causes of the destruction of so many works of the elder nations. In metaphysical abstractions these changes, this destruction of material forms, is referred to Time; and the Unknown gives his view of the operation of Time, philosophically considered. He sees that the principle of change is a principle of life; that without decay there can be no reproduction; that everything belonging to the earth, whether in its primitive state or modified by human hands, is submitted to certain and immutable laws of destruction, as permanent and universal as those which produce the planetary motions. Man may conserve from ruin only for a period:

Yet, when all is done, that can be done, in the work of conservation, it is only producing a difference in the degree of duration. And from the statement that our friend has made, it is evident that none of the works of a mortal being can be eternal, as none of the combinations of a limited intellect can be infinite. The operations of nature, when slow, are no less sure; however man may, for a time, usurp dominion over her, she is certain of recovering her empire. He converts her rocks, her stones, her trees, into forms of palaces, houses and ships; he employs the metals found in the bosom of the earth as instruments of power, and the sands and clays which constitute its surface as ornaments and resources of luxury; he imprisons air by water, and tortures water by fire to change or modify or destroy the natural forms of things. But, in some lustrums, his works begin to change, and in a few centuries they decay and are in ruins; and, his mighty temples, framed as it were for immortal and divine purposes, and his bridges formed of granite and ribbed with iron, and his walls for defence, and the splendid monuments by which he has endeavoured to give eternity even to his perishable remains, are gradually destroyed; and these structures, which have resisted the waves of the ocean, the tempests of the sky, and the stroke of lightning, shall yield to the operation of the dews of heaven, of frost, rain, vapour and imperceptible atmospheric influences; and, as the worm devours the lineaments of mortal beauty, so the lichens and the moss, and the most insignificant plants shall feed upon his columns and his pyramids, and the most humble and insignificant insects shall undermine and sap the foundations of his colossal works, and make their habitations among the ruin of his palaces, and the falling seats of his earthly glory.

The Sixth Dialogue was not intended to conclude the book; Davy had planned to write a Dialogue on the doctrine of Definite Proportions – of which a very short example is printed in the fifth volume of the *Works* – and to follow it by a Dialogue on the Chemical Elements. A fragment of this Dialogue is included in the ninth volume of the *Works*. Other editions of *Consolations in Travel* end with the Sixth Dialogue, a good resting-place for the imagination, but not intended as Davy's own.

In his will Davy desired that the proceeds of the book – like *Salmonia* it was published by John Murray – should go towards the education of his godson, Humphry Millett, eldest son of his sister Elizabeth (Betsy) who had married her second cousin, Lieutenant

John Boulderson Millett. Kitty and Grace did not marry. They lived together near the house where they were born. Kitty, it is rumoured, grew a little formidable and witchlike as she grew older; she paced Penzance carrying always a large umbrella.

19 · *The Last Journey*

Consolations in Travel was written during Davy's last continental journey. He left England on March 29th, 1828, accompanied by his godson, John Tobin, son of his old friend, James Webbe Tobin, who had died on his plantation at Nevis in 1814.

John Tobin had been a medical student at Heidelberg; he acted as Davy's secretary. Although Davy, in letters to his wife, sometimes complained of the young man, calling him his *wildeman*, and referring to his laziness and slovenly ways, he seems on the whole to have been well served by him as well as by his servant, George, who had more of his confidence. He could not have been an easy travelling companion. We catch glimpses of him in his great cloak lined with white fur; we hear of the eighty books he took with him from England – the number of books made Thomas Hood say in later years that Davy's consolation had been not in travel but in a library. He had his dogs, his guns, and his fishing tackle, though 'Fished in vain' is a not infrequent entry in his diary. In the absence of friends he required that Tobin should be constantly reading aloud to him. He mocked at his youthful companion's medical advice given with a novice's confidence, and scandalized him by trying various remedies from apothecaries' shops. Tobin, after Davy's death, wrote an account of his travels with the philosopher; but he adds little to the knowledge we have from other sources.

He and Davy travelled by way of France and Belgium to the Rhine, and from the Rhine to the Danube. The late spring, summer and early autumn were spent in various parts of Illyria and Styria with, once again, Laibach (now Ljubljana) as the centre from which Davy made various excursions. He described in *Consolations in Travel* and in letters to Lady Davy the Illyrian maid who, more than anyone else, refreshed him in his weakness and weariness. She was Josephine or Papina Dettela, a daughter of the innkeeper. She had nursed him on one of his previous visits, and her constant attendance and kindness had dispelled his melancholy. In *Consolations in Travel* he makes her the figure of a dream. In spite of the sceptical looks of his friends he relates to them how, when he was suffering from gaol-fever – obviously his illness of 1807 – he had had a constant vision of a beautiful creature with light brown hair and blue eyes and how, twenty years later, he had found her in person at Laibach. Lady Davy must have teased him in a letter about the *Bettinas* of his earlier days; for he tells her in his reply about his little friend and nurse at Laibach. The weather and his Papina seem to have combined in graciousness to him. He wrote in far better spirits than usual to Lady Davy on October 5th, 1828:

My dear Jane – I have received all your letters, and last the two from Trieste. I have remained here for many reasons, but I shall go to Trieste as soon as I hear from Mr During that he has procured a live torpedo for my experiments. The first time since my illness, I have found a month pass too quickly here. The weather has been delightful, and I have had enough shooting to make the day short; and my pursuits in natural history respecting the migration of birds, have given me some new and curious results. I must not forget the constant attention and kindness of my 'Illyrian maid', I mean poetically and really. The art of living happy is, I believe, the art of being agreeably deluded; and faith in all things is superior to reason, which, after all, is but a dead weight in advanced life, though as the pendulum to the clock in youth.

I am delighted with your account of your improvement in your health, which I trust will be permanent. I have certainly improved more since I have been here than at any former period of my illness, and I *sometimes forget* my miseries.

Do as you please *about the house*. I have now resolved to winter in Italy, though the *ubi* is still uncertain: for want of a sure address, after I leave this, write to me at Bologna.

Mr W. T.'s, I dare say, is an agreeable house. Vanity is always an agreeable quality, and is, I think, the most exquisite and odorous essence of selfishness, and almost always connected with good nature, and, when the stomach is right, with good temper.

God bless you, my dear Jane! I am just going to look for quails.

Your affectionate,

H. DAVY.

Not only Papina but the people of the Austrian Tyrol generally, as he had made clear in his fishing journal, moved him to admiration. He was attracted by the innocence of feeling and gaiety of the Italian and Austrian peasants who associated religion with their high festival humours. He praised the open churches, the freedom of the poorest to approach unhindered, contrasting this freedom with the locked doors of London churches. Even Westminster Abbey was at that time, in general, closed. Blake contrasted the bitter churches with the alehouse so pleasant and warm. There was, Davy said, a personal freedom, an absence of social segregation in Italian and Austrian religion which, though differently based from the legal freedom of the individual citizens in his own land, was as precious. With his own extreme activity of intellect he had had, in his youth, what Wordsworth called a correspondent hurry of delighted feeling. As he grew older he came to think that it was not curious knowledge, but nurture of the religious instincts, and practice of the human pieties which made for earthly happiness. Sometimes he sounds almost cynical in his letters to Lady Davy, as though any illusion were better than utter disenchantment. But his last letters are the letters of a man made capricious by pain; they often express contradictory sentiments and opinions.

John, Archduke of Austria, had met Davy on a previous occasion, and had written affectionately to him. He had given him every facility for pursuing knowledge and sport in the loosely held provinces, and had invited him to be his guest. The Crown Princes of Sweden and Denmark had done no less; but Davy was more temperamentally akin to the Catholic south than to the Protestant north. He was an ardent supporter of Catholic emancipation at home, in this disagreeing with his admired Robert Peel who, until his whirlabout, when he introduced the Catholic Relief Bill, had himself opposed it. Davy was in no sense a party man; he might have been called a Tory-Whig, not reactionary, sympathetic to reasonable reform, but never extreme, and distrusting all doctrinaire notions of easy happiness for clustered humanity through re-organization of the cluster while the individual souls remained selfish and indifferent. Without being a sharp or pitiless judge of human nature he had met stupidity, malice, and prejudice; and he had begun to know himself. It was not religious enthusiasm – though he told Lady Davy he was becoming 'more apostolic' as he grew older – which made him support the Bill. It was a feeling that there was neither peace nor security for England without it. When he heard that the Bill was carried he wrote to Lady Davy, 'I rejoice that the Catholic question is carried. Without having a strong political bias, I have always considered this point as essential to the welfare of England as a great country, and connected with her glory as a liberal, philosophical and Christian country.'

His letter rejoicing in the passing of the Bill was written to Lady Davy on March 1st, 1829. The interval between November and March had been spent by him in Rome, dictating *Consolations in Travel* to Tobin, riding out a little, conversing a little with Spada and Morichini, sometimes hopeful of recovery, sometimes despairing. It was during one of his more hopeful moods that he wrote his last letter to Tom Poole. At that time he wished that Poole might join him for a spring or summer tour of his favourite regions. He wrote:

Rome, February 6th, 1829.

My dear Poole,

I have not written to you during my absence from England, because I had no satisfactory account of any marked progress towards health to give you, and the feelings of an anvalid are painful enough for himself, and should, I think, never form part of his correspondence; for they are not diminished by the conviction that they are felt by others. Would I were better! I would then write to you an agreeable letter from this glorious city; but I am here *wearing away* the winter; a ruin among ruins! I am anxious to hear from you – very anxious, so pray write to me with this address, 'Sir H. Davy, Inglese, posta restanti, Rovigo, Italia.' You know you must pay the postage to the frontier, otherwise the letters, like one a friend sent to me, will go back to you. Pray be so good as to be particular in the direction – the 'Inglese' is necessary. I hope you got a copy of my little trifle *Salmonia*. I ordered copies to be sent to you, to Mr W— and to Mr Baker: but as the course of letters in foreign countries is uncertain, I am not sure you received them; if not, you will have lost little; a *second edition* will soon be out, which will be in every respect more worthy of your perusal, being, I think, twice (not saying much for it) as entertaining and philosophical. I will take care by early orders that you have this book. I write and philosophize a good deal, and have nearly finished a work with a higher aim than the little book I speak of above, and which I shall dedicate to you. It contains the essence of my philosophical

[244]

opinions, and some of my poetical reveries. It is, like the *Salmonia*, an amusement of my sickness; but *paulo majora canamus*. I sometimes think of the lines of Waller, and seem to feel their truth:

> The soul's dark cottage, batter'd and decay'd
> Lets in new light through chinks that Time has made.

I have, notwithstanding my infirmities, attended to scientific subjects whenever it was in my power, and I have sent the Royal Society a paper which they will publish, on the peculiar Electricity of the Torpedo, which I think bears remotely on the functions of life. I attend a good deal to Natural History, and I think I have recognized in the Mediterranean a *new species of eel*, a sort of link between the conger and the muraena of the ancients. I have no doubt Mr Baker is right about the distinction between the conger and the common eel. I am very anxious to know what he thinks about their generation. Pray get from him a distinct opinion on this subject. I am at this moment getting the eels *in the markets here* dissected and have found *ova* in plenty. Pray tell me particularly what Mr Baker has done; this is a favourite subject with me, and you can give me no news so interesting. My dear friend, I shall never forget your kindness to me. You, with one other person, have given me the little happiness I have enjoyed since my severe visitation.

I fight against sickness and fate, believing I have still duties to perform, and that even my illness is connected in some way with my being made useful to my fellow-creatures. I have this conviction full on my mind, that intellectual beings spring from the same breath of Infinite Intelligence, and return to it again, but by different courses. Like rivers, born amidst the clouds of heaven, and lost in the deep eternal ocean – some in youth, rapid and short-lived torrents; some in manhood, powerful and copious rivers; and some in age, by a winding and slow course, half lost in that career, and making their exit through many shallow and sandy mouths. I hope to be at Rovigo about the first week in April. I travel slowly and with my own horses. If you will come and join me there, I can give you a place in a comfortable carriage, and can show you the most glorious country in Europe – Illyria and Styria, and take you to the French frontier before the beginning of autumn – perhaps to

England. If you can come, do so at once. I have two servants and can accommodate you with everything. I think of taking some baths before I return, in Upper Austria; but I write as if I were a strong man, when I am like a pendulum, as it were, swinging between death and life.

God bless you, my dear Poole,

Your grateful and affectionate friend,

H. DAVY.

Pray remember me to our friends at Stowey.

Too much concentration on *Consolations in Travel*, and on preparations for experiments, took toll of his failing strength. It was after dictating to Tobin an addition to the Sixth Dialogue of the *Consolations* that, when trying to rise and go to his bedroom, he found that he had lost all control of his limbs, although his mind was clear. He might write to Lady Davy that the search for knowledge did not make for human happiness; but his own pursuit of knowledge terminated only with his life. His letter to John in which he bequeaths to him his final scientific investigation and urges him to continue it expresses an inextinguishable curiosity about life which no recantation made in moments of weariness and pain could cancel. The letter was written by Tobin at Davy's dictation:

My dear John,

If it had not been for this attack my intention was to have gone to Fiumicino or Civita Vecchia to make some experiments on the torpedo. I hope you will take up this subject, which, both as a comparative anatomist and chemist, you are very capable to elucidate. You will see my paper on the torpedo in the manuscript book, which I have left in Mr Tobin's hands. It was my wish to have exposed an unmagnetised needle to the continued shocks of a torpedo in a metallic spiral, making the metallic communication perfect with both electrical organs. There is in my little box an apparatus for this purpose, which I hope you will use. Large living torpedos may be procured at Fiumicino or Civita Vecchia. The shock from a very small jar will make a needle magnetic, provided it is entirely passed through the metallic conductors; but I did not find this effect when there was any interruption by water. There are many things worth attending to in the two kinds of torpedinal fishes found here – the tremula and ochiatella. Pray do not neglect this subject, which I leave to you as another legacy. God bless you, my dear brother!

Your affectionate friend,

H. DAVY.

In a postscript he wrote with his own hand, 'My dear John', but no more. Tobin completed the postscript from dictation, 'I have written to you, but I fear you have not got the letter. I have this moment received your address. I am dying. Come as quickly as you can. You will not see me alive, I am afraid. God bless you.'

John Davy received this letter at Malta where he was stationed as Physician to the Forces. It took six days, although he was permitted to travel in Vice-Admiral Sir Pultenay Malcolm's tender, to reach Naples; from Naples he set out with a courier at midnight and arrived in Rome early in the morning of February 16th. He was directed by Dr Morichini to his brother's lodgings.

Davy had at this time a conviction that he would not recover; but he had lost the irritability of temper which had accompanied his illness. He showed gladness at his brother's presence, was cheerful and tranquil, and his mind was very clear. For the greater part of the day John sat by his bedside reading *Consolations in Travel* aloud, and discussing difficulties as they arose. Next morning he went to market for torpedoes which he dissected in the room next to Humphry's. He was employed alternately in reading to his brother, and in carrying out the dissection, occasionally showing the results, or reading aloud notes of what he had done. He says that Humphry was not only amused but deeply interested.

But as the days passed Humphry's lassitude increased, and when the critical reading of *Consolations in Travel* ended, he seemed to be without motive for exertion, and considered this the appropriate time to die. John had been reading to him the first half of Moore's *The Epicurean*. 'The sad colouring', he said, 'and melancholy sentiment which pervade that elegant little work, with the wildness of some parts of the fiction, and its marvellous subterranean scenery and incidents, pleased him much.' When night came Humphry would not allow his brother to remain in his room, not even on a couch as he had done before. He was sure he would die that night, and was in the mood of Byron's *Euthanasia* which perfectly agreed, he said, with his own sentiments. His mind was so disposed for death that when John went to him on the following morning and drew back the curtains, he expressed great astonishment at being alive. He said he had gone through the whole process of dying, and that when he awoke he had difficulty in convincing himself that he was in his earthly existence; he was under the necessity of making certain movements to satisfy his mind that he was still in the body: raising the hand to intercept the light, lifting the bed-clothes, closing his eyelids.

He lived to feel and smell and hear another April and May in the countries he loved. For he began to grow stronger. On the 26th March he wanted no more of *The Epicurean;* it was too melancholy, he said. He had to be read to all the time, but now he chose his old favourites: Shakespeare, *The Arabian Nights,* and *Humphry Clinker* which, in its beginning, transported him to Hotwells. Lady Davy, summoned from England, arrived early in April. The weather was delicious; the rains were past; the sky was clear and blue over the green Campagna. It is affecting to read John Davy's account of his brother at this time. During the drives in the neighbourhood of Rome the spring seemed to renew the dying man as it renewed the flowers and leaves. Davy loved to be refreshed by the air. He and John went to revisit the milky stream of the Albula lake described in the Third Dialogue of *Consolations in Travel,* or they drove towards Albano to marvel again at the ruins of Rome; or they chose Humphry's favourite way to Civitavecchia, either keeping close to the Tiber or at a little distance from the course of the river, through the beautiful hilly region which stretches in that direction – 'a succession of gardens, vineyards and orchards, villas and farm-houses, reminding one of England; the orchards in full bloom, and the vineyards rapidly sprouting and bursting into leaf, under the influence of the warm sun and air'. Four or five miles out, between the hills, in a little sheltered hollow rich in wild flowers, Davy would practise walking, or sit in the open air listening to the larks.

On April 20th he sent his last note to Penzance. John Davy told him he was writing home, and Humphry wished to add a word. He wrote:

My dear Sister,

I am very ill, but thanks to my dearest John, still alive. God bless you all.

H. DAVY.

He held the pen with difficulty because of his physical weakness and the paralysis in his hand. Yet he continued to grow stronger. He was able to go out in a carriage and see the splendid illumination of St Peter's on Easter Monday night. Then, for coolness, a journey to Geneva was proposed.

The party, consisting of Humphry, John and their attendants, set out for Geneva at the end of April. Lady Davy went ahead to make comfortable arrangements at each stage. From April 30th to May 28th they drove, not day after day, but pausing to rest where the inns were good – two days at Siena, two at Florence, six at Genoa (where, because of a serious relapse on Davy's part, the project was nearly abandoned), two days at Turin and one at Susa. On May 23rd they crossed Mont Cenis and slept that night at Lansleburg. Then, travelling on successive days, they came early on May 28th to Geneva.

During this journey, Davy's passion for natural history still burned; he retained his relish of light mountain air, his pleasure in diverse scenery, his eye for sport when his hand had failed him, and even his memory of the delicate savours of food. Although John Davy's style tends to be opaque and repetitive he lets the world in at many windows during this last journey. We catch sight of the embayed and wooded shores of the Mediterranean as the brothers drive by the newly constructed sea and mountain road towards Genoa; the rich profusion of gardens and groves round the ramparts of the city; the freshness of everything after heavy rain on the way to Turin; the contrast of the Alpine summits and the bright red vetch in a narrow valley; the mules drawing the carriage in an easy manner over the Mont Cenis Pass, where John remembered the winter horror he and Humphry had experienced on a former occasion when travelling to Ravenna; the bits of thistledown floating in the air which the sunshine lit and made to seem like insects. Davy's contention that bodily pain could be met by abstracting the mind was partly borne out in this last month. Perhaps Jane Davy was, after all, a better wife for Humphry than is generally supposed. She left him alone when he wanted to be alone, and she did not oppose this last journey and make him dwindle to death in a house. Wordsworth's loving wife would have been horrified had William gone off to hear again the sounding cataract when he was sick to death.

On May 28th the party reached Geneva and put up at La Couronne. They were at the inn early in the day. Humphry lay resting on a sofa; but sometimes he would go to a window, look at the lake, and long once again to throw a fly. He dined at four o'clock, enjoying the fish provided, concerned about its cookery as he had been at Mullion, when he tried to make the sauce for the landlady. He joked with the waiter, and told him he wanted to taste every kind of fish the lake provided before he left. But when Lady Davy told him of the death of Dr Thomas Young he was very much moved; Dr Wollaston had died not many weeks before. He went to bed and, in the early morning before it was light, he died. John had been called and was with him. He died, leaving neither son nor daughter; but he had known what it was like to feel immortal. Now rest, after a life of intense activity, was in a favourite phrase of his, agreeable with the analogy of things.

When Wordsworth heard of his death, at a time when he was grieving at Scott's seizure and the continued illness of Coleridge, he wrote in a letter to Rogers, 'Davy is gone. Surely these [the story-teller, the poet and the scientist] are men of power, not to be replaced should they disappear, as one alas has done.'

Davy was only midway through his fifty-first year when he died. He had lived faster than most people. Lady Davy lived to be seventy-five. In the memoirs and letters of the period there are many notices of her during her widowhood; some of her sayings and doings have coloured Davy's reputation as well as her own. Among those who attended the funeral at Geneva was the author of *The History of the Italian Republics*, Sismondi, who had married Jessie Allen, a sister of Josiah Wedgwood's wife. The sisters thought Lady Davy a clever and brilliant person, and liked her in spite of her association with some of whom they disapproved. But Jessie Sismondi, when she met Lady Davy about a month after Davy's death, renewed the friendship with a certain reluctance. She wrote to her niece, Emma Wedgwood, from Chêne on July 9th:

> Lady Davy came a few hours after our return. I suffered her to go away to her inn without inviting her to return to us. I fancied I saw she was disappointed; and it is painful to disappoint people's expectation of you, and I felt uneasy; and yesterday when we dined with her at her inn and saw that she was melancholy, solitary and nervous, I pressed her to return to us, and she comes on Saturday to

breakfast. If the weather permits we are going another little tour with her.

Much of Lady Davy's later life was spent in making little tours, and in recalling stories of her former travels with her husband. She loved to act as guide to younger people. We see her glittering now in her house in Park Street; now at Richmond; now in Belgium; now in Geneva; now in her favourite Rome; journeying much, though Sydney Smith had said as early as 1814 that she knew the roads of earth as well as the wrinkles on Rogers' face.

Her talents were not negligible. A woman well able to judge of them was Mary Somerville, one of the two or three women of her period learned in science. She was Jane's contemporary, and a fellow-Scotswoman. To read of Mary Somerville's heroic efforts to prevent her mind from being stifled is to be filled with wonder at such power of persistence. When Mary's sailor father heard that she was trying to learn Euclid he said to her mother, 'Peg, we must put a stop to this, or we shall have Mary in a strait-jacket one of these days. There was X who went raving mad about longitude.' Mary Somerville won her way to intellectual eminence as a mathematician and scientist, and to the full enjoyment of her unusual powers, through every kind of obstacle. Her testimony to Jane Davy's accomplishments, and also to her habit of showing off on her weakest side, is of interest because it is given sympathetically. The two ladies were in Rome where, long after Davy's death, Lady Davy was playing an active part in society. Mary Somerville said of her:

> She talked a great deal and talked well when she spoke English, but like many of us had more pretension with regard to the things she could not do well than to those she could. She was a Latin scholar, and as far as reading and knowing the literature of modern languages went she was very accomplished, but unfortunately she fancied she spoke them perfectly, and was never happier than when she had people of different nations dining with her, each of whom she addressed in his own language. Many amusing mistakes of hers in speaking Italian were current in both Roman and English circles.

Lady Davy continued to spar with Rogers. Her favourite Lord Dudley, whose eccentricities, mingled with wit, learning, and sound judgment, made such a fascinating passage in the story of her friendships, had reached madness and death. She attracted amusing men. John Hookham Frere liked her. She wrote to him often, making Lady Cadogan jealous when they were all wintering in Rome in the autumn of 1842. 'I am made very unhappy', Lady Cadogan wrote to him,[1] 'by the crowing of Lady Davy who comes to me at least once a week with messages from you – thereby letting me know that you write to *her*, and not to me . . . some mutual friends of yours and the lady's insinuate that she does *not* hear from you *quite* as often as she says she does.' Lady Cadogan was irritated by Jane's parade of grander friendships. She told Frere, 'I have not seen our *Jane* to wreak my woman's vengeance by saying you had written to me, but I shall not miss the opportunity whenever it may occur – but Jane does not cultivate *us*. She is always talking (and I suppose) thinking of The Carlisles, The Grosvenors, The Sunderlands, Lady Grenville and Lady Granville – and so on. She gives Tea and Dinners but not to The Cadogans, altho' she comes here whenever I ask her.'

As a widow, Jane never had quite the prestige which had been hers when she made her wedding tour with Davy in Scotland. She began many of her stories with, 'Once, when I was staying at Dunrobin'.

Sydney Smith's notices of her continue to vary with his mood, and with the person he is writing to. To his wife he said, after dining with 'the Davy', that she had been 'noisy, flattering and full of her dreadful reasoning eloquence'. To John Allen he named her *Davy de Medici*, so much did she love to play the cicerone in the world of Italian art and history. But he knew her kindness to be always available, especially to the young and unsure; he guessed at her large charities. In 1841 he wrote her a letter which throws light on him and on her and on the times so changed since the days when she had been *a dasher* and a *bel esprit*, and he in high fooling. He wrote from Combe Florey:

Combe Florey   Taunton,
August 31, 1841.

My dear Lady Davy,
  I thank you for your very kind letter which gave Mrs Sydney and me much pleasure and carried us back agreeably into past times. We are both tolerably well, bulging out like old houses but with no immediate intention of tumbling down.

---

[1] *John Hookham Frere and His Friends*, Gabrielle Festing, pp. 244–9.

The Last Journey

The country is in a state of political transition and the Shabby are preparing their consciences and opinions for attack. The Queen likes it least of anybody, and declares loudly that the coming Prince of Wales she is sure will resemble Peel in countenance and manner. A short prayer against this tucked into the Litany might possibly do some good.

I think all our common friends are doing well. Some are fatter, some more spare, none handsomer, but such as they are I think you will see them all again – but pray do you mean to see any of us again or do you mean to end your days in Rome? a town I am told you have entirely enslaved, and where in spite of your protestantism you are omnipotent. – Your *Protestantism* but I confess that reflection makes me sometimes melancholy; your attachment to the clergy generally, the activity of your mind, the Roman Catholic spirit of proselytism, all alarm me. I am afraid they will get hold of you and we shall lose you from the Church of England. Only promise me that you will not give up till you have subjected your arguments to my examination and given me a chance to reply. Tell them there is a Canonico dottissimo to whom you have pledged your theological faith. Excuse my zeal, it is an additional proof of my affection, which wants however no additional proof. God bless you, my dear Lady Davy.

Your affectionate friend,

SYDNEY SMITH.

Evidently his arguments prevailed. She died a member of the Church of England and received a temperate notice in the *Gentleman's Magazine* as 'the relict of a Baronet'.

Her life had been very largely a performance; her letters betray her. But had Jane Austen made her customary inquiry as to this lady's principles, she would have received a reassuring answer. Jane Davy's principles were not pretence. Humphry trusted her honour, generosity, and sense of justice. He made her the sole executrix of his will, though his brother, sisters, and nephew were the principal legatees.

John Davy thought it would have been better both for Humphry and Jane had they never met; but we must allow in John for some prejudice. He lived far into the Victorian Age, not dying until 1868. He, too, married a Scot. She was Margaret Fletcher, daughter of the marriage between Archibald Fletcher, an advocate of Edinburgh, an ardent supporter of reform, with Eliza Dawson, whose autobiography, published under her married name, Mrs Fletcher of Lancrigg, is still a delightful book to read. Of Margaret and John Davy's children, two survived them, Archibald and Grace. John and his wife had settled at Lasketh-How, not far from Wordsworth – who went to Court wearing Humphry Davy's sword which he borrowed either from John or from Lady Davy – an event which called forth Benjamin Haydon's pained, indignant letter to Wordsworth that a former republican should so fall from grace.

As inspector-general of army hospitals John Davy undertook various government missions. Abreast of current knowledge in his profession, humane and hardworking, an able scientist, he was an enlightened man to advise those in authority on affairs of administration, combining independent criticism of rulers with a sympathy for the ruled rare in any day; but he entirely lacked Humphry's unpredictable fire. Humphry was a strange Davy; as Michael was a strange Faraday, Cavendish a strange Cavendish, John the painter a strange Constable, and S. T. Coleridge a most strange issue of the parsonage at Ottery St Mary.

As for Humphry, when all the official eulogies had been pronounced, and the elegiac verses written – Davies Gilbert's and Baron Cuvier's were the best of the one, Valentine Le Grice's and William Sotheby's of the other – he paid the usual penalty for having been the height of fashion. He had begun to pay it before he died. Considered on the surface in a worldly way, his is a success story ending in an anticlimax; considered as an allegory, it is the progress of an inquiring spirit towards humility. Davy came to know enough to know and confess that he knew nothing; so he was not dogmatic either in science or religion. Gentle towards other people's superstitions and prejudices and, strangely enough, a more rare virtue as one looks back through time, towards their varied pleasures, he remained, in spite of Wordsworth's fears for him, what Wordsworth wished every experimental philosopher might also be, a dear and genuine inmate of the household of man. Perhaps this was because he retained, from his childhood in West Penwith, deepening to awe as his exploration grew more profound, his faculty of wonder and his impulse to adore.

The Youth who Carried a Light

I saw him pass as the new day dawned,
  Murmuring some musical phrase;
Horses were drinking and floundering in the pond,
  And the tired stars thinned their gaze;
Yet these were not the spectacles at all that he conned,
  But an inner one, giving out rays.

Such was the thing in his eye, walking there,
  The very and visible thing,
A close light, displacing the gray of the morning air,
  And the tokens that the dark was taking wing;
And was it not the radiance of a purpose rare
  That might ripe to its accomplishing?

What became of that light? I wonder still its fate!
  Was it quenched ere its full apogee?
Did it struggle frail and frailer to a beam emaciate?
  Did it thrive till matured in verity?
Or did it travel on, to be a new young dreamer's freight,
  And thence on infinitely?

THOMAS HARDY
Moments of Vision, 1915.

A Select Bibliography

ABBOTT, C. C., 'The Parents of T. L. Beddoes', *Durham University Journal*, June 1942, pp. 159–75

ALLEN, WILLIAM, *Life of William Allen*, with Selections from his Correspondence, 3 vols, 1846

ATKINS, ANNA, *Memoir of John George Children* (printed privately), Westminster, 1853

ATKINSON, A. D., 'The Royal Society and English Vocabulary', *Notes and Records of the Royal Society of London*, vol. 12, no. 1, pp. 40–3

BABBAGE, CHARLES, *Reflections on the Decline of Science in England*, and some of its Causes, 1830

——, *Passages from the Life of a Philosopher*, 1864

BLUNDEN, EDMUND, *Shelley, A Life Story*, Collins, 1946

BOWEN, JOHN, *A Brief Memoir . . . of William Baker*, 1854

CARDEW, ALEXANDER, *Cornelius Cardew*, 1926

CARLYON, C., *Early Years and Late Recollections*, 2 vols, 1843

CARTWRIGHT, F. F., *The English Pioneers of Anaesthesia*, J. Wright & Sons, 1952

CLARKE, ISABEL C., *Maria Edgeworth: Her Family and Friends*, Hutchinson, 1950

COLERIDGE, S. T., *Collected Letters of Samuel Taylor Coleridge*, 4 vols, 1785–1819, ed. E. L. Griggs, Oxford University Press, 1956–9

——, *The Notebooks of Samuel Taylor Coleridge*, 1794–1804, 2 vols, ed. Kathleen Coburn, Routledge, 1957

COTTLE, JOSEPH, *Reminiscences of S. T. Coleridge and Robert Southey*, 1847

CROSSE, CORNELIA, *Memorials of Andrew Crosse*, 1857

CUVIER, BARON GEORGES, *Éloges Historiques*, 3 vols, 1861

DARWIN, EMMA, *Family Letters, 1792–1896*, 2 vols, Murray, 1915

DAWSON, WARREN R. (ed.), *The Banks Letters: A Calendar of the manuscript correspondence of Sir Joseph Banks preserved in the British Museum, and other collections in Great Britain*, 1958

A Select Bibliography

DAVY, SIR HUMPHRY, *The Collected Works*, edited by his brother John Davy, 9 vols, 1839, with the *Memoirs of the Life of Sir Humphry Davy* as vol. i

DAVY, JOHN, *Memoirs of the Life of Sir Humphry Davy*, 2 vols, 1836

——, *Fragmentary Remains Literary and Scientific*; with a sketch of his life, and selections from his correspondence, 1858

——, A letter from John Davy, to the editors of the *Philosophical Magazine*, in reply to certain charges made by C. Babbage against the late Sir Humphry Davy, when President of the Royal Society, 1864

DINGLE, HERBERT, *The Scientific Adventure*, Pitman, 1952

FARADAY, MICHAEL, *Faraday's Diary*, publ. under editorial supervision of T. Martin, 7 vols, Bell, 1932–6

FESTING, GABRIELLE, *John Hookham Frere and His Friends*, 1899

FOX, HENRY EDWARD (afterwards fourth and last Lord Holland) *The Journal of the Hon. Henry Edward Fox 1818–1830*, ed. the Earl of Ilchester

GRABO, CARL, *A Newton Among Poets*: Shelley's Use of Science in *Prometheus Unbound*, Oxford University Press, 1930

GREGORY, J. C., *The Scientific Achievements of Sir H. Davy*, Oxford University Press, 1930

GREVILLE, C. C., *Journals of the Reigns of George IV and William IV*, ed. Henry Reeve, 3 vols, 1874

HILLARD, GEORGE STILLMAN (ed.), *Life, Letters and Journals of George Ticknor*, 2 vols, 1876

INGLIS-JONES, ELIZABETH, *The Great Maria*, Faber, 1959

JONES, JOHN, *The Egotistical Sublime*, 1954

KENDALL, JAMES PICKERING, *Humphry Davy: 'Pilot' of Penzance*, Faber, 1954

METEYARD, ELIZA, *A Group of Englishmen*, 1871

MOORE, THOMAS, *Memoirs, Journal and Correspondence of Thomas Moore*, ed. Lord John Russell, 8 vols, 1853

PARIS, JOHN AYRTON, *The Life of Sir Humphry Davy, Bart.*, 2 vols, 1831

REDDING, CYRUS, *Fifty Years' Recollections, Literary and Personal*, 3 vols, 1856

REES, EDGAR A., *Old Penzance*, 1956

SANDFORD, MRS HENRY, *Thomas Poole and His Friends*, 2 vols, 1888

SHARROCK, ROGER, 'The Chemist and the Poet: Sir Humphry Davy and the Preface to the Lyrical Ballads', *Notes and Records of the Royal Society of London*, vol. 17, no. 1.

SOMERVILLE, MARTHA, *Personal Recollections from early life to old age of Mary Somerville*, 1873

SOUTHEY, ROBERT, *Common Place Book*, ed. J. W. Warner, 4 vols, 1949–51

STOCK, J. E., *Memoirs of the Life of Thomas Beddoes*, 1811

THOMSON, BENJAMIN, COUNT VON RUMFORD, *Essays, Political, Economical and Philosophical*, 4 vols, 1796–1802

THORPE, T. E., *Humphry Davy, poet and philosopher*, 1896

TOBIN, J. J., *Journal of a Tour made in Styria, Carniola, and Italy, whilst accompanying the late Sir Humphry Davy*, 1832

TYNDALL, JOHN, *Faraday as a Discoverer*, 1868, 1870

UNDERWOOD, THOMAS RICHARD, *Journal of a Détenu*, 1825

WARD, JOHN WILLIAM, FIRST EARL OF DUDLEY, *Letters to Ivy*, ed. S. H. Romilly, 1905

——, *Letters of the Earl of Dudley to the Bishop of Llandaff*, 1840

WEIL, E., An Unpublished letter by Davy on the Safety Lamp, *Annals of Science*, vol. 6, no. 3, March 1950, p. 306

WRIGHT, THOMAS, and EVANS, R. H. (eds.), *Account of the Caricatures of James Gillray*, 1851

Report from Select Committee on Accidents in Mines, together with Minutes of Evidence and Index, 4th Sept., 1835

Index

Index

Index

of Wordsworth, 63–4; on Recovery from a Dangerous Illness, 97; after the death of Byron, 183; on training the young, 203

Lectures: introductory to a course of chemists, 84; Geology and Mineralogy of Cornwall, 89; chemical agencies of electricity, 93; volcanoes, 154; ancient Greek and Roman pigments, 155; the safety lamp, 165; the papyri of Herculaneum, 183; electromagnetism, 201; electricity and chemical changes, 201, 207; application of liquids formed by condensation of gases, 205

Books: early essays on heat, light and respiration, 35–9; *Researches* into nitrous oxide, 41–6; *Elements of Chemical Philosophy,* 129–31; *Elements of Agricultural Chemistry,* 129, 132, 139; *Salmonia,* analysed, 222–9; *Consolations in Travel,* analysed, 230–40; mentioned, 51, 52, 152, 154, 178, 219–20, 241–2, 244, 247

Davy, Jane (Lady): her family, 118; courted by Davy, 122–6; marriage, 126–9; social life in London, 132–3; accompanies Davy to Continent, 142–61; contemporary (1815–18) opinions of her, 175–7; as wife of the P.R.S., 198–200; as housekeeper, 215; relations with Scott, 215–17; care for Davy in illness, 229, 248–50; widowhood, 250–3; mentioned, 219, 242

Davy, John: childhood, 8, 11, 15, 16, 17; on John Tonkin, 14; on Humphry's appearance, 28; education, 69–72; works with Humphry in London, 99; studies in Edinburgh, hears of Jane, 126, 128–9; Humphry's care of him, 131–2; becomes army doctor, 158, 162; accompanies the sick Humphry to Italy, 217–18; care of him in last ill-

ness, 246–50; subsequent career 253–4; mentioned, 19, 21, 51, 133–4, 184, 204, 208, 213

Davy, Katherine, 8, 17, 70, 100, 214, 240

Davy, Robert (father), 5–6, 7, 9, 17

Davy, Robert (great-uncle), 5

Dawson, Eliza, 254

De La Mare, Walter, 160–1

De la Roche, M., 136

de Quincey, Thomas, 111

Dettela, Papina, 241–2

Devonshire, Duchess of, 34

Dudley, Earl of (J. W. Ward), 127, 131, 134, 199–200, 219, 251–2

Dugast, M., 16, 70–1

Dulong, 133

Edgeworth family, 32, 114–15, 121

Edgeworth, Maria, 114–15, 132–3, 198, 199

Electromagnetic rotation, 201–3

Evans, R. H., 79

Evolution, 226–7

Faraday, James, 134

Faraday, Margaret, 134–6

Faraday, Michael: upbringing, 134–6; application to Davy and appointment to Royal Institution, 136–9; accompanies Davy on the Continent, 142–5, 150, 151, 152, 155–7; promotion at the Institution and first lectures, 162; on the safety lamp, 174; Davy's interest in him, 180; troubled election to Royal Society, 201–4; mentioned, 206

Faraday, Robert, 136

Felling colliery, 163

Festing, Gabrielle, 252(n)

Flanders, 178–9

Fletcher, Margaret, 253

Florence, 151–2

Fontainebleau, 150

Fox, Henry Edward, 127, 198

French Revolution, 2, 16

Index

Index